Tools for Modeling Systems: Learning the Language of Patterns

Peter R. Bergethon

Symmetry Learning Press

Symmetry Learning Systems and Press
5 Bretton Road
Dover, MA 02030
www.symmetrylearning.com

SYMMETRY LEARNING
PRESS
A subsidiary of
SYMMETRY RESEARCH, INC

Symmetry Learning Systems and Symmetry Learning
Press are wholly owned by Symmetry Research, Inc., Dover, MA 02030

9 8 7 6 5 4 3 2 1

ISBN 978-1-58447-100-4 Tools for Modeling Systems: Learning the Language of Patterns

Contents

Note to this Edition

This volume, **Tools for Modeling Systems: Learning the Language of Patterns**, (ISBN 978-1-58447-100-4) is a modified second edition of the previous text "Learning the Language of Patterns: A Teacher's Guided Tour". This edition, now completely revised, uses the same approach to the Language of Patterns and the Graphical Organizer system developed for the SymmetryScience program, the topic of the original edition. SymmetryScience is a unified science education program that is designed to start in pre-school years and includes materials available electronically for K-9 years. The SymmetryScience program is designed for use by parents, classroom teachers, home school programs or supplementary science education programs. A new second edition for the SymmetryScience program containing the earlier content from "A Teacher's Guided Tour" is also now available for parents, teachers and science educators who are using SymmetryScience. This volume is entitled, Tools for Critical Thinking: Learning the Language of Patterns, (ISBN 978-1-58447-102-8).

Tools for Modeling Systems: Learning the Language of Patterns is designed to be used independently from the SymmetryScience curriculum. It is targeted for the growing community of people interested in learning to use systems analysis and scientific modeling in their work and study. In response to requests from this community of learners this edition introducing The Language of Patterns as a stepping stone to competence in systems analysis and systems modeling is now offered. More materials and information on this approach to the practice of scientific inquiry are available from the retailer of this volume and at www.symmetrylearning.com.

"Science is a way of looking at the world" Happy looking!

Peter R. Bergethon   January 2013

CHAPTER 1

Introducing the Progression of Inquiry and A Language for Modeling Systems

The Age of Knowledge

The begininning of the twenty-first century heralded the turn from one remarkable era into one that is more complex and even more exciting. This is the turn from the Information Era to the Age of Knowledge. The amount of information and the rate at which it is created, dispensed, and revised is mind-boggling. The success of our civilization depends on our capacity to acquire and use this torrent of information in effective and meaningful ways.

The effective and meaningful use of information is called **knowledge**. Knowledge is formed from information. The construction of knowledge requires an analytical process to knit the information into useful, information-rich models that can be used in our everyday lives.

Since we are interested in the use of available information and not just its collection, we must strive to move into the Age of Knowledge and not just live in the Information Age. That is the goal of the process we will call **The Progression of Inquiry**.

What is a Model?

A key idea in the Progression of Inquiry is a recognition of the central role played by **modeling** in human understanding. Modeling is the process of representing something (called a system) using a partial accounting of its features and properties. Modeling is a *simplifying* process. This means that the model is an abstraction (a simplification) of the thing under investigation. A model, even though it is a partial representation of the real world, is acted on and treated as if it is the original. A "good" model provides a satisfactory representation *for the purposes under consideration*.

Often modeling is confusing at the start. This is because sub-models are used to build up a larger model of the particular thing of interest. The result is that there will be a series of abstract representations and symbols used to represent a complicated system. It takes tools and practice to keep track of the parts of a system and its series of representations. The Language of Patterns is one of these tools.

A model is a system description that includes identifying the parts, relationships, context and characteristics of that system under observation. The Language of Patterns and the graphical organizers that are the subject of this book will help you learn how to make these descriptions.

Modeling and Building Knowledge

Modeling is the coordinating link between observation and understanding. It is central to the practice of the scientific method. Interestingly, models are also how the brain represents the world that it regards. Our brains extract certain features of the world. The brain uses these pattern features to construct a series of interrelated internal models at several levels:

- **Descriptive or representational models** - describe what parts there are and how they go together.
- **Explanatory models** - describe the cause and effect relationships between the parts in the system observed.
- **Predictive models** - used by the brain to predict and plan for future events and actions.

These models form a progressive series moving toward increasingly abstract relationships. In the human mind, such a series of internal mental models represent the movement from "simple" experience to expert knowledge and understanding of our world.

Human brains are not naturally scientific model makers	The human progression of inquiry always involves description and explanation. As we will see in later chapters, scientific modeling requires the addition of experimental models with strict rules of verification. The brain does not require this additional level of modeling in order to accept or use an explanation. It is important to be aware of this limitation since assumptions, beliefs and opinion can often be substitutedin models and mistaken for more objective, verifiable truths. This can lead to substantial error in the modeling process. The scientific method uses specific steps to verify models. In the remainder of this book we will be concerned with scientific models and the scientific progression of inquiry.
Modeling in Science	Scientists, applying modern scientific methods of inquiry, use a parallel series of model-building steps to learn about, organize, and explain events in the natural world. Scientists

- observe phenomena and make **descriptive models** that define the features and pattern elements,

- then try to explain the phenomenon by creating an **explanatory model**. The desciptive model is re-organized into an explanatory or theoretical model. This is accomplished by posing **hypotheses** or proposed relationships between the system parts. These relationships are proposals for connecting how one part influences the others and ultimately the overall output properties of the system.

- A unique aspect of scientific thinking is that there must be a careful testing of the proposed explanatory relationships (hypotheses) by using **experimental models** to verify the explanatory model against evidence collected from the real or natural system.

As evidence accumulates to support the hypothesized relationships, a scientific theory is evolved. A **theory** is a model of current understanding and explanations that results from the interplay of observation, descriptive modeling, explanatory modeling, and experimental evidence.

These models (along with their elements, assumptions, and relationships) are the "content" known as scientific knowledge and relayed in science education. Scientists use the theoretical models as a starting point to ask questions and explore natural phenomena to a deeper degree. As new scientific information (observations and measurements) are made, the explanatory or theoretical model is either confirmed or changed to accommodate the new information. Thus the process and content of the progression of inquiry to produce scientific knowledge is intertwined in the enterprise of science.

The Progression of Inquiry Requires a Language

Modern science and the critical-thinking practices that it supports depends on these three linked stages of model making.

A flow diagram can be used to graphically relate these steps of knowledge construction.

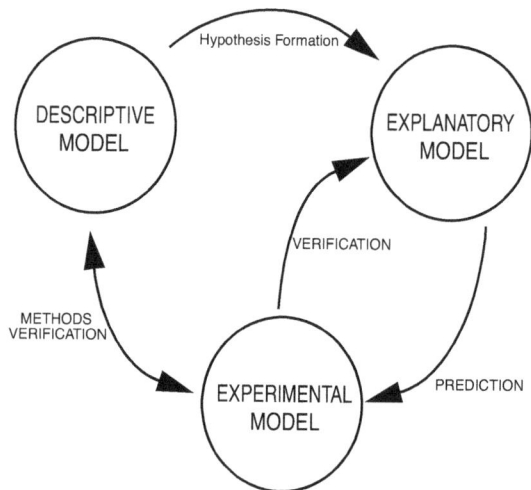

The construction of these models requires a systematic language that render a model of reality. Systems are described by recognizing that they contain definable **elements** that can be **related** to one another and appear in the **context** of a particular background space or dimension. The interaction of the parts with one another and their context space gives rise to system

properties that are often distinct from the individual characteristics of the components of the system. The language that captures systems information is the subject of this book, **"The Language of Patterns"**. It might also be called the "Language of Models". Use of this language is a key approach and tool needed for mastering the process of critical thinking.

The Language of Patterns

The Language of Patterns is a descriptive modeling language. It is used at each step of the knowledge construction process or Progression of Inquiry. When an observation is made, the Language of Patterns expresses the elements of the observation in terms of their properties, their arrangement, and the background space or context in which they are arranged.

Any object or system of objects can be fully characterized by a description of its pattern features, which include its emergent **properties** that result from the interaction of the system **elements**, **rules**, and **background** space. Using these fundamental ideas, the Language of Patterns develops the tools for critical thinking and analysis that underlie all knowledge construction. This method of learning is simple, yet powerful enough to help understand rocket science, medicine, electronics as well as economics, and ice hockey play-making! Despite its capacity to analyze complex topics, the method is so easy to use that a kindergarten class can begin it at the start of the school year. Once mastered, the Language of Patterns can be used over and over to characterize, explore, and understand any system of interest at multiple levels.

Scientific inquiry uses the Language of Patterns at every level from observation, to model building, to hypothesis testing, to experimental verification. Critical thinking skills and the process of scientific inquiry are practiced anytime the Language of Patterns is used, anywhere, for anything. When a model is built, the Language of Patterns allows symbols, such as chemical, technical, and medical notations as well as letters, musical notes, and mathematical equations to be assigned to each element, arrangement, and background space.

As the need for the logical tools of deduction and inference develop in the modeling process, the Language of Patterns is the method of choice to extract the rules and connections from the constructed model.

During the **verification process**, a hypothesis is posed which can be proven or disproven by collecting evidence. The Language of Patterns relates the observation to the model, to the hypothesis, and to this gathering and interpretation of the experimental data.

Finally, when the knowledge construction process calls for opening the discovery process to outside evaluation, the Language of Patterns provides a common, precise language of communication and discussion.

The Language of Patterns is structured just as scientific inquiry is structured. Therefore:

- Teaching the Language of Patterns is teaching science.
- Using the Language of Patterns is doing science.
- Teaching with the Language of Patterns is teaching with science.

A series of graphical organizers, the Graphical Analyzer System, assist all users to become fluent in and apply the Language of Patterns throughout their studies and work.

Unit One

The Language of Patterns

Notes

CHAPTER 2

Getting Started with the Language of Patterns

The Language of Patterns Vocabulary

The Language of Patterns is extremely useful throughout the process of knowledge construction and critical thinking. It enables the use of modeling which is the key tool that promotes effective problem solving. It is an extremely powerful yet simple method that enhances critical thinking and fosters creativity.

The Language of Patterns has six major ideas, which work together.

These are:

- Systems
- Properties
- Elements
- Rules
- Background Space (or Context)
- Evolution (or Change)

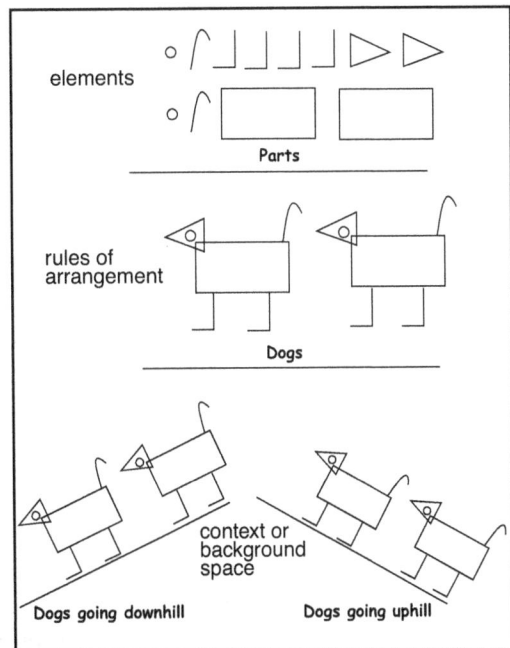

elements

Parts

rules of arrangement

Dogs

context or background space

Dogs going downhill   Dogs going uphill

The Language of Patterns is used to describe the construction of information and knowledge in terms of the pattern features. These pattern features are the elements, rules, and background space of an object, idea, or system that give rise to system properties.

- **Systems** are characterized by **properties** (color, shape, texture, size, weight, states).
- System properties emerge from the interaction of system elements governed by certain **rules** in a **background** or **context space**.

Systems result when elements are put together.

- Every system is composed of a certain group of parts or events that we will generally call **elements**.
- In a system, the elements are arranged somewhere - in what is called a **background space**.
- The **arrangement** of elements is governed by **rules of relationship** between each element **and** between the elements and the background space.

This is an example of a system:

- The **elements** of this system are seven objects each of whom is a *subsystem* whose **properties** are that they are squares with varied fill patterns.
- The **background space** in which they are arranged is a flat page [in technical terms: a plane of two dimensions (length and width)].
- One of the **rules** that arranges the squares is that their center-points lie in a straight line.

Properties and Observables

We can describe each system in terms of its own **properties**. A property is a characteristic of a system (structure, object, event) that describes it. By comparing and contrasting the properties or attributes of different systems, we can group and sort them. These are fundamental critical analysis skills. It is important to recognize that the elements that compose a system are themselves subsystems and have properties that also allow them (the subsystem) to be grouped or sorted.

Properties form the basis for sorting, grouping, comparing, contrasting, and categorizing things.

- To fully describe an system, every property would have to be listed and enumerated.
- This is usually impossible and unnecessary since most systems have many, many properties. We hardly ever have to know or name every property in order to be able to usefully sort, contrast, or compare systems.

When we work with systems, we choose a property that we can easily observe or measure. The particular properties that are chosen to observe or measure are called **observables**.

An important part of critical thinking and analysis is to be able to specifically state what the observables are that are being used to characterize a system.

- The choice of an observable is almost always an assumption. Careful skeptical critical inquiry must always question that choice.
- **CAUTION**: The choice of observables can turn out to be the limit, prejudice, or error in the analysis of a system.

The following example demonstrates the advantage of picking the correct observable. Consider this system of objects:

If we need to know how many chairs to set up in a classroom so that everyone can have a seat, the proper observable would be the number of people. Having the properties of their names, favorite color, food, baseball team, religion, and income might be interesting, but it would not help us set up the room.

The problem of picking the wrong observable can also be demonstrated by using the same system of objects:

This group of people is going to vote for President of the United States. The correct observable is whether they are citizens of the USA and are registered to vote. If we tested them for the ability to read and used that to sort them into voting and non-voting groups, in 2012 this would be against the law. [The Civil Rights Act of 1964 made it illegal to use a test of literacy as the observable to decide whether a citizen of the US could vote or not.]

Analyzing Systems

If the elements in the system example are grouped by the observable of their fill pattern, the system can be analyzed as shown in Figure A. Systems can be described by their properties as shown in Figure B:

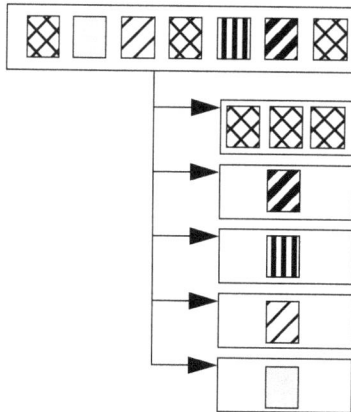

repeating pattern
straight (or linear)
rectangular
silent
unmoving

Figure A Fill Pattern

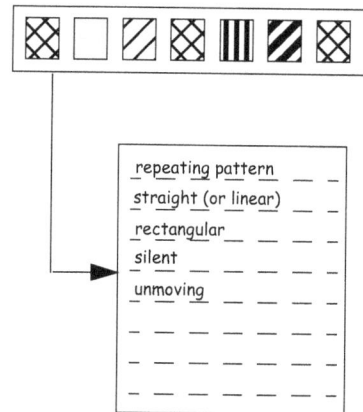

Figure B Properties

When we want to describe a system, we need to describe the elements in it, the background space where the elements are arranged, the rules that arrange the elements in the space, and the properties that emerge from all of these system interactions.

When we look at our world, it is the systems that we notice. Cars are systems, people are systems, poems are systems, songs are systems. The package of elements, rules of arrangement, and background space gives rise to the characteristics (observable properties) that make us laugh at jokes, cry at tragedies, fall in love, see the stars twinkle, and navigate by the constellations.

The Language of Patterns gives us the critical analysis tools to explore systems and their properties, yet it reaches its full power when turned to the exploration of the most compelling and interesting aspects of our existence, how the world **changes** and **evolves**. These dynamic processes occur when structures or systems are moved to action. It is the dynamic analysis of change and evolution for which the Language of Patterns was designed. We will learn how this important task is accomplished in the coming pages.

Notes

CHAPTER 3

The Language of Patterns -
The Basis of Critical Thinking

Knowledge Construction through Critical Thinking

The Language of Patterns is necessary throughout the process of knowledge construction and critical thinking. Knowledge is formed when a higher level model is formed from new information. This new information is incorporated into previously known knowledge. Higher level models are constructed by arranging the new system elements into new model structures.

Acquisition of Information

This process can be conceived as being derived from three domains: the **Experienced World**, the **World of Critical Analysis**, and from the **Modeled World**.

1.   The **Experienced World** includes both the **Natural World** and the **Imaginative World**.

The Experienced World includes experiences such as:
- The taste of an apple,
- Feeling the wind in your face,
- The rush of falling in love,
- Seeing the stars on a clear night.

Experience is gained when information is acquired through observation. The Experienced World can be discovered without education. It is detailed, specific, diverse, and non-uniform.

2.   The **World of Critical Analysis** results from the output models of critical analysis and analytical thinking. When writers, musicians, performers, or scientists express what they experience or how they feel about the Experienced World, they make a model of the experience using the craft of critical analysis.

The world of critical analysis is:

- An author's use of syntax, grammar, and style,
- A theatre director's choreography and stage directions,
- An architect's drafting methods,
- A musician's notes, staff, and orchestrations,
- A scientist's mathematical and chemical notations.

The critical analysis process describes and measures a system in terms of pattern features, then maps it onto a the abstractions that comprise the model.

3. The **Modeled World** is a "knowledge structure" created when the information is mapped into a **knowledge model**.

Examples of knowledge structures are:

- A written language and the canon of literature,
- A dramatic and dance production,
- An architect's drawings,
- A musician's symphony,
- A scientist's theory of the universe.

Knowledge models are representations of how we understand the world. When we regard the natural/real world, we extract certain observable features through our senses. These features are then translated into a model that exists only in symbolic form. Such systems of representation include:

- The symbolic languages of mathematics,
- Spoken and written language,
- Musical composition,
- Chemical notation,
- Graphical art, and others.

Knowledge systems are unified, abstract, and coded. Education is required to provide the codes. The mapping of experience onto a model system and communication with others require **critical thinking** strategies and skills. These skills are refined with education. A formal education teaches the vocabulary, syntax, and use of these systems.

This diagram summarizes the three Worlds which are the sources of information for the brain and mind:

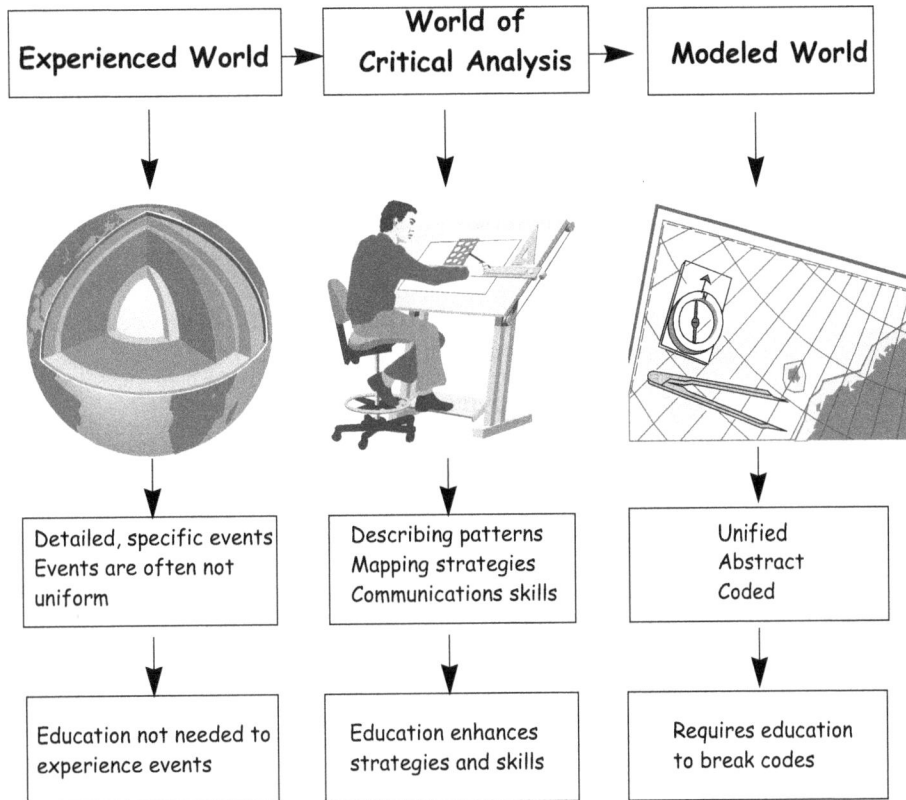

| Experienced World | → | World of Critical Analysis | → | Modeled World |

| Detailed, specific events Events are often not uniform | Describing patterns Mapping strategies Communications skills | Unified Abstract Coded |

| Education not needed to experience events | Education enhances strategies and skills | Requires education to break codes |

Five Steps to Modeling and Critical Analysis

The process of acquiring knowledge about systems through modeling and critical analysis may be organized into five steps. A simple diagram can guide this discussion.

Step 1. Observation

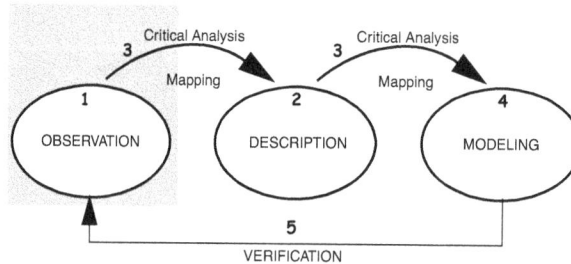

Observations are made in the natural world by the human senses. Tools, such as a microscope or weight scale, are often used to extend the senses.

The pattern features of observed objects and events (the properties of systems) are detected (observables).

Step 2. Description

The description of pattern features is an important critical thinking skill because it enables recording, discussion, and debate with others about the

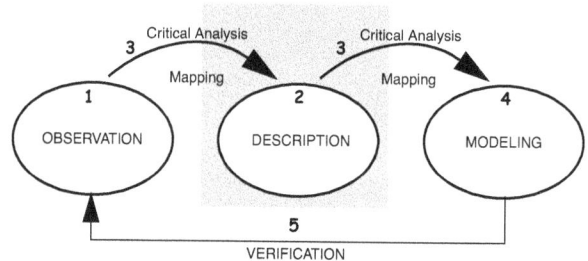

observation. The Language of Patterns is used to describe the pattern features of the system in terms of its properties, elements, rules that relate the elements to each other, and background space in which the system being observed exists.

In the background our brain is performing pattern feature extraction and building internal brain models simultaneously and continuously. Often we may not be aware or conscious of this process. When the process and its models are brought to mind and subjected to analysis, a critical thought process is begun.

Step 3. Critical Analysis

Mapping the described features onto a model system requires a critical analysis of the observations.

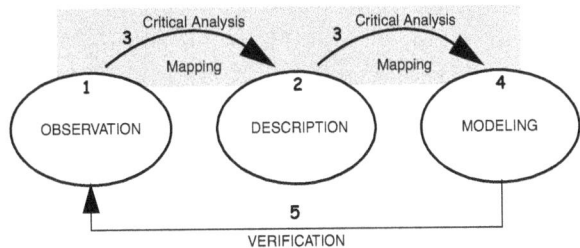

- The **observables** of the system describe what properties can be measured and how a set of properties define a state or condition of a system. The particular values of those properties allow categorization of the system.

- How the **elements** will be represented in the model must be determined. For example, will words, latin phrases, musical notes, chemical notation, mathematical symbols, or colors be used to symbolize each element in the system?

- The characteristics of the **background space** in which the mapping will be formed is necessary. Usually the background space has the same number of dimensions but may be scaled larger or smaller. However, as the pictured globes show, this does not have to be the case.

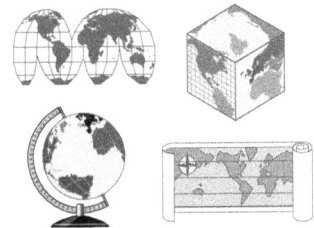

- A definition of the **rules** that arrange the elements of the structure being described into the background space is required. Elements can be ordered:

  - <u>By naming and counting (nominal ordering)</u> - who and how much there is.

  - <u>By ordinal measure (ordinal ordering)</u> - arranging the elements by height, weight, or in the order that timed events happen.

  - <u>By arrangement in space (spatial ordering)</u>, e.g., 30° north latitude and 15° west longitude.

Critical analysis is the aspect of knowledge construction that depends on logic, rational analysis, precision in language, principles of mathematical thinking (including sorting, ordering, categorizing, transforming), and discovering ideas of number, form, and difference. The skills of critical analysis are the mapping skills used to build a model of some aspect of the world.

Step 4. Modeling

Once the mapping process is completed, a model exists. A model is a representation of the world that exists only as an abstraction. It is created by the

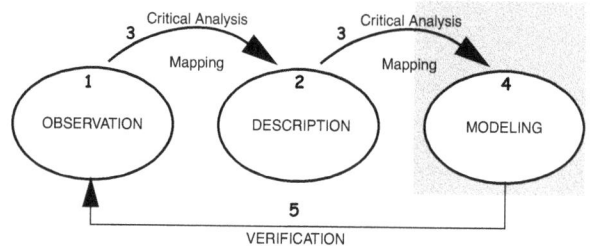

process of pattern feature extraction and critical analysis.

Step 5. Verification

The methods of verification and error checking in a critical mode of thinking require a willingness to define and formally represent an idea or observation,

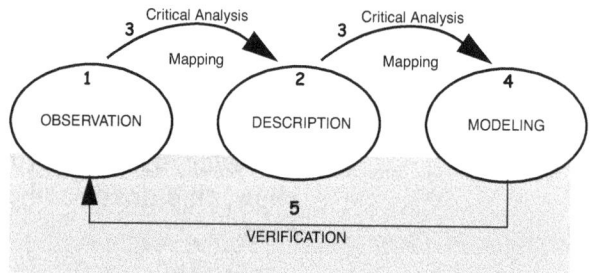

communication, and acceptance of another's critical analysis of an idea.

This approach to modeling works for descriptive, explanatory or experimental models and is used repeatedly in the Progression of Inquiry. The following diagram demonstrates the dynamic interaction of the Language of Patterns and the construction of knowledge models. The stars indicate each point where the Language of Patterns is used.

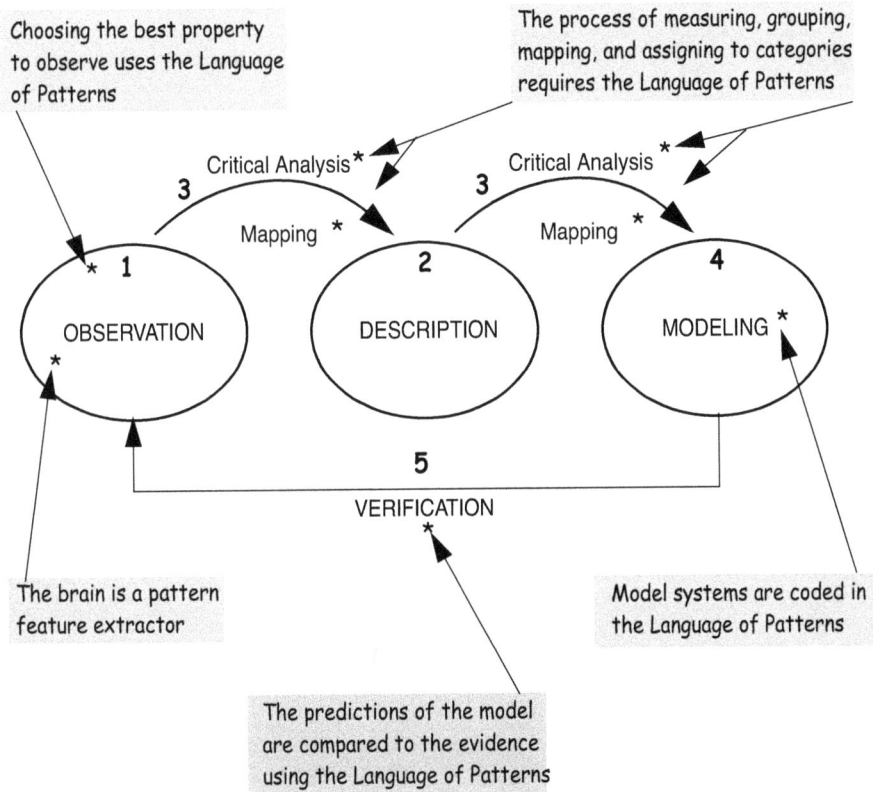

Choosing the best property to observe uses the Language of Patterns

The process of measuring, grouping, mapping, and assigning to categories requires the Language of Patterns

Critical Analysis *

3

Critical Analysis *

3

Mapping *

Mapping *

* 1

2

4

OBSERVATION

DESCRIPTION

MODELING *

*

5

VERIFICATION
*

The brain is a pattern feature extractor

Model systems are coded in the Language of Patterns

The predictions of the model are compared to the evidence using the Language of Patterns

The Language of Patterns helps form the foundation of a literacy in critical thinking and analysis. It helps us describe some aspect of our world and lets us relate that particular part to all the other parts of our world. The critical analysis tool reflected in the Language of Patterns is part of a fundamental literacy that includes literacy in science, mathematics, and the arts.

Notes

CHAPTER 4

The Language of Patterns and the Scientific Method

The Scientific Method

Science is a particular way to look at and explore the world. It is a habit of mind for building knowledge. Modern civilization is built on the idea that the best method for advancing knowledge is through the human endeavor of experimental and critical science. Modern science has evolved over the last four centuries to become the highest form of critical exploration with its careful methods of verification and the evaluation of ideas based on data and measurement.

The scientific method is an organized way of exploring and describing the world and proceeds in a series of steps. The characteristics of the scientific method are:

- Observation.
- Making models and proposing hypotheses.
- Testing and retesting by experimentation.

There are two sub-divisions to the scientific process, one **theoretical** and the other **experimental**.

- In theoretical methods, observations about the world are organized into a model of the world.
- In experimental methods, the theoretical model is tested by making observations of an experimental model.

- In the scientific method, the theoretical model is compared to the experimental model.

  - 👍 If they agree, the experiment supports the theoretical model.

  - 👎 If they disagree, the experiment rejects the theoretical model.

Science Begins With Observation

All science begins with observations of the patterns of the natural world. A knowledge of the patterns of natural events and the resulting effects are among the most basic interests of all people. Since the dawn of our species, survival has depended upon recognizing these patterns which have included day-night cycles, the motions of the sun and moon, the turn of the seasons, the migration of animals, the rhythm of human existence.

Archeological evidence from cave paintings and the notching of bone and reindeer horns suggests that pre-historic humans were extremely careful in their recording of seasonal and temporal patterns. Such knowledge is acquired by simple observation. However, observation alone is not "modern" science. Observation is "proto-science". Proto-science accepts observations without question or verification.

Skepticism in Modern Science

In modern science, what is known is viewed skeptically. A **skeptical** view is characterized by the continual questioning of the certainty with which something is considered "true". "How sure am I, that I actually know what I think I know" is the skeptical view. This includes a skepticism of the ability of the human to objectively make observations. The scientific methods of error-checking use the experimental examination of a natural system to verify what is known. Modern scientific thinking requires proof of how cause and effect are connected and also that correlations are not just due to chance.

Modeling is the Core of the Scientific Method

Relating a cause to an effect is the purpose of a model. In a model, a particular observable (a cause) is linked to another particular observable (the effect). The cause is an observable that an experimenter will vary. The effect is an observable that the experimenter has theorized to be related to the cause. A working model explains how things happen. Investigating cause and effect uses logic, inference, and deduction.

In modern science, the fundamental rules of the model are discovered from careful consideration, measurement, experiment, and analysis of simple cases. Observation is the first step in constructing a descriptive model. The measured attributes first are mapped onto a descriptive model then later a theoretical model. The models are usually made from a "formal system" of equations, words, or geometric figures.

The validity of the model is tested by:

- Making a prediction based on the hypothesis.
- Performing experiments in the natural world to test the model.
- Making experimental measurements.
- Recognizing that the observer may influence the experiment and measuring that influence.
- Changing the model when the experimental evidence requires modification.

Modern Science is Based on Evidence

The explanatory cause-effect model is constantly modified with experimental evidence that verifies or falsifies the ideas on which the model is constructed. In experimental science, the relationship between the cause-observable and the effect-observable is called a **hypothesis**.

- A hypothesis is not just a question - it is a specific kind of question.

- A hypothesis states: **The cause is related to the effect in the following way. Is this relationship true or false?**

Verification in Modern Science

In order to verify with some certainty the validity of a hypothesis, experimental-skeptical methods include:

- **Control** experiments to be certain that the effect observed is related to the proposed cause.
- **Calibration** of measurements to be sure that an agreed upon standard is used to quantitatively compare measured effects between experiments and experimenters.
- **Context** of the system under study with reference to other systems.
- **Communication** between experimenters to confirm and evaluate each other's experimental and theoretical work.

This overall process is the Modern Scientific Method.

The Progression of Inquiry is the Modern Scientific Method and our two diagrams can be seen to be the same with a strict set of validation rules added:

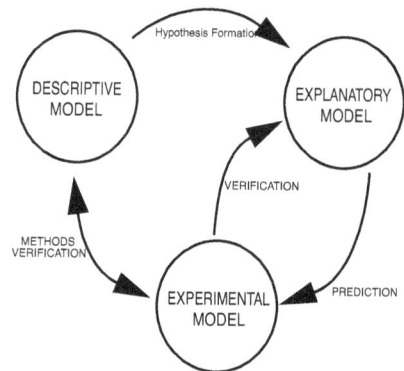

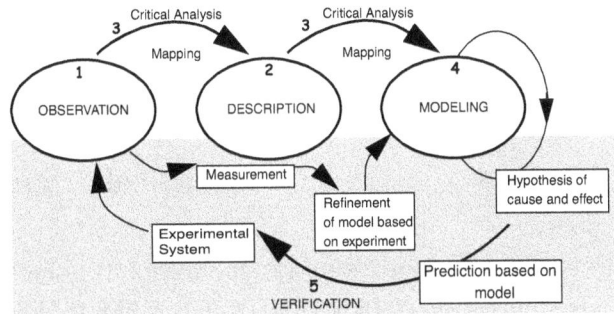

The Scientific Method and the Progression of Inquiry

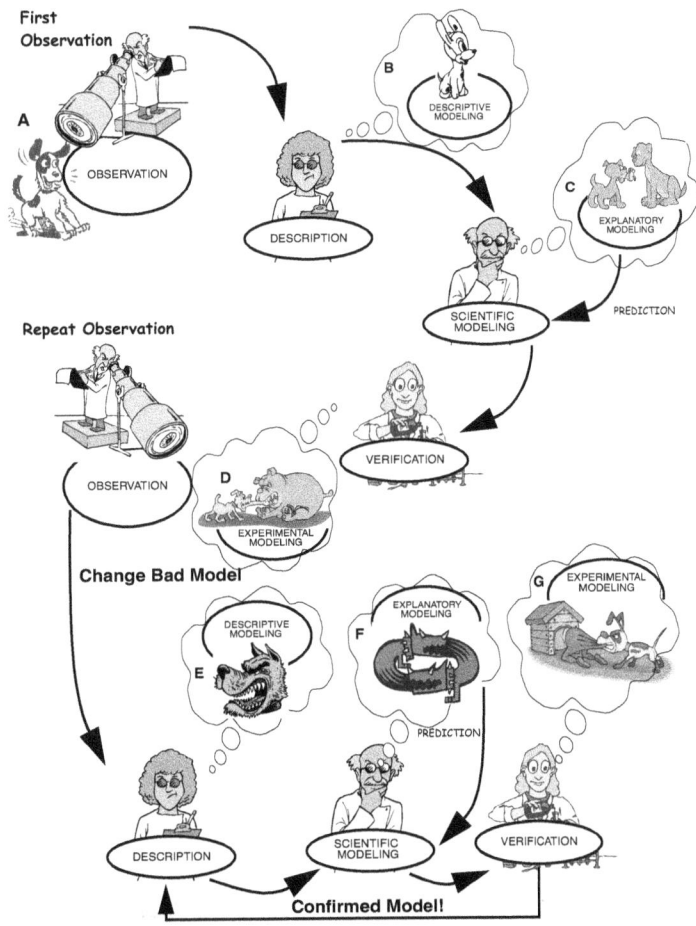

Scientific
Method

In order to emphasize the process, the steps used in the practice of science are now reviewed. The following description includes more detail on the two subdivisions of the Scientific Method, theoretical and experimental branches.

Theoretical Steps

1.   **Observations** are made about some part of our world. For example, we may notice that the sky is blue. The observation is described by a Language of Pattern analysis. Scientific observation accounts for and measures:

- The properties of the systems,
- The elements or parts of the system,
- The way the elements are arranged,
- The background space where the elements are arranged, and
- The background conditions or properties under which a particular arrangement is found.

2.   A **descriptive model** of the observations is made. Symbols represent the observed and measured elements. Examples of symbols are:

- words,
- mathematical expressions,
- musical notes and writing,
- dance steps,
- ultural rituals,
- electronic schematics, and
- chemical element notations.

3.   **Cause and effect** create explanatory models. Reasoned hypotheses are made of the relationships of the system. These rules tell us how one element or rule (a cause) will affect a systems property (the effect). A model that explains how system

properties emerge from the relationship of elements, rules of arrangement, and the background space is called a **theoretical model**.

Experimental Steps

In the scientific method, the theoretical model must always be tested by **experiment**. **Evidence** in the scientific method comes from careful **experimental design**. The method of experimental science is described as a series of steps.

1. **Define the problem to be explored.**
    *   Usually a specific question is asked that focuses attention on one aspect of a model. This focus is often labelled the **problem** that is under investigation.
    *   For example: a scientist might be interested in the problem: "Where do animals get oxygen?"

2. Make a hypothesis.
    *   A **hypothesis** is a special kind of question. It proposes a relationship between two parts of a system.
    *   The relationship can be between elements, elements and the background space, the rules of arrangement and the elements, or even different rules.
    *   A good hypothesis proposes, "If I make this change in this part, I will be able to observe that effect on that part".
    *   The relationship proposed by the hypothesis leads the experimenter to make a prediction. A **prediction** is the logical outcome of the hypothesis and how the model works.
    *   A good hypothesis offers a single proposition that can be shown to be true or false by experiment.

· With a good hypothesis, an experiment can be designed that specifically tests to see if the proposed relationship actually exists.

· A conclusion is reached that the proposed relationship exists if the prediction matches the actual observed result from the experiment.

3. Design an experimental system to test the hypothesis.

· An experimental system is a working model of the real world. However, it is a simplified or **controlled environment**.

· A controlled system is one in which every aspect of the system can be defined and described. Therefore, it can be repeated by someone else at another place and time.

· Experiments generate data to explore how one property in the system (the **cause**) is related to a second property (the **effect**). Experimenters obtain this data by making and recording measurements.

· Both of these properties, the cause and the effect, must be able to be observed and measured. They are special experimental properties called **observables**.

· In an experiment, the measured values of observables can change. Observables that change or vary their value are usually called **variables**.

· If the experimenter varies or changes the value of one of the observables, this is called the **independent variable** (the cause).

· The variable whose value is changed by the independent variable is called the **dependent variable** (the effect).

4. Gather data by measurement and observation from the experimental system.

· Data are collected by measuring and observing the variables in the experiment.

- The relationship between the cause and effect properties is determined by analyzing the data.
- Experimental results are compared to a **control experiment**.
    - Scientists are skeptical of their own experiments, therefore every experimental system has a **control experiment**.
    - A control experiment is an experimental set-up that is used as a **standard** or **reference** for the system.
- This means that everything in the control and experimental set-up are the same except that the independent variable (the cause) is taken out of the control experiment.
- If the dependent variable (the effect) changes in the control experiment, the changes in the experimental system can not be attributed to the independent variable (the "cause" was never there!).
- Importantly, experimental systems all contain degrees of uncertainty: uncertainty in measurement, variability in the natural response of the system to produce an observable measurement. These must be taken into account.

5. Draw a conclusion that supports or rejects the hypothesis.
    - If the experimental data gives the same result as that predicted by the hypothesis within the degree of uncertainty, then a conclusion can be made that the hypothesis is supported.

6. Repeat the experiment; have other scientists repeat it.
    - Since the scientific method is skeptical, the experiment is done many times by many different scientists. If the same conclusion is reached by everyone, then the hypothesis is generally accepted as a fact by the scientific community.

- As more and more of the ideas that went into building the model are accepted as facts, the model may be accepted as a **scientific theory**.

- A **scientific law** is different from a scientific theory. If after many, many tests by many, many scientists in many different situations, the relationship proposed by a hypothesis is always proved correct, it may be called a scientific law.

As a result of the success of the scientific method in exploring and understanding the natural world, the scientific model of investigation has become the *de facto* standard for virtually all forms of critical human inquiry. This includes fields of art, language, and social science.

Unit Two

The Graphical Analyzer System

Notes

CHAPTER 5

Graphical Analyzers and System Modeling

Weaving the Strands of Knowledge

Mastery requires time, work, and motivation; progress is made by the continuous and diligent application of all three. Understanding comes from knowledge, which relies upon information. The scientific method of inquiry is the tool that when mastered fosters the free movement from information to understanding. This process is guided by the progression of inquiry. The Language of Patterns, through the Graphical Analyzer System, is a map to this "road of inquiry".

The Graphical Analyzer System

This Unit will introduce and guide your preparation for using the Graphical Analyzer System.

In practice, the Graphical Analyzers are intended to structure how anyone can use observation, description, measurement, modeling, and experimental design to learn and practice science inquiry.

- Using the Analyzers continuously and progressively through learning provides a way to practice the common threads of scientific inquiry and knowledge construction.

- Each Analyzer is introduced and demonstrated in this Unit. Examples showing its application to learning science and critical thinking are explored.

The Organization of the Graphic Analyzer System

The Graphic Analyzer System is organized around the vocabulary of the Language of Patterns. There is a variety of related graphic analyzers that support the basic ideas used in the Language of Patterns:

1. Systems

2. Properties

3. Elements

4. Rules

5. Background

6. Change

As recalled in the diagram below, the Language of Patterns is closely linked to these steps in the scientific and critical thinking process and are guided by the Progression of Inquiry's successive models.

1. Observation
2. Descriptive Modeling
3. Critical Analysis and Hypothesis Mapping
4. Explanatory Modeling
5. Verification by prediction and testing in an experimental model.

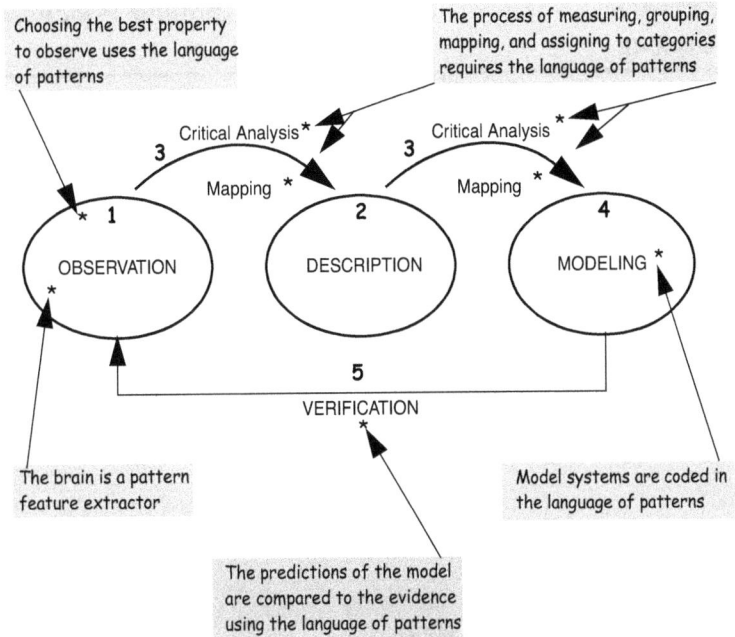

Choosing the best property
to observe uses the language
of patterns

The process of measuring, grouping,
mapping, and assigning to categories
requires the language of patterns

Critical Analysis *

Critical Analysis *

3

3

Mapping *

Mapping *

* 1

2

4

OBSERVATION

DESCRIPTION

MODELING *

*

5

VERIFICATION
*

The brain is a pattern
feature extractor

Model systems are coded in
the language of patterns

The predictions of the model
are compared to the evidence
using the language of patterns

The following chart indicates the Graphical Analyzers which are used with the five basic concepts of the Language of Patterns. The critical thinking skills and the scientific inquiry process associated with each Graphical Analyzer is listed.

Language of Patterns Concept	Critical Thinking Skill	Scientific Inquiry Process	Graphical Analyzer
Systems Elements	Properties Observables	Observation Description	Element Analyzer Element Counter Element Extractor Property Analyzer
	Sorting Classifying	Observation Description Mapping	Sorting Analyzer (single observable) Sorting Analyzer (multiple observable) Sub-sorting Analyzer Classifying Analyzer Compare and Contrast Analyzer Conflict Analyzer
	Describing Mapping	Description Mapping Modeling	Technical Writing Analyzer Coding Grid
Background Space	Properties	Observation Description Modeling	Background Space Analyzer Boundary Analyzer
	Measuring Mapping	Description Modeling Verification	Background Feature Extractor

Language of Patterns Concept	Critical Thinking Skill	Scientific Inquiry Process	Graphical Analyzer
Rules	Arrangements	Observation Description Mapping	Arrangement Analyzer
Rules	Inferring rules of arrangement	Modeling Verification	Rule Extractor
Structure	Describing	Observation Description Mapping Modeling	Structure Analyzer Modified Structure Analyzer
Evolution or Change	Describing	Observation Description Modeling	Evolution Analyzer
	Predicting	Modeling Verification	Evolution Analyzer

Using the Graphical Analyzer System

We are now ready to examine the specific Graphical Analyzers in the System.

- The Graphical Analyzers are described along with examples to demonstrate each Analyzer.
- Practice Exercises will include Analyzers that show how science content can be organized. Occasionally, the use of an Analyzer across knowledge domains will be demonstrated by giving examples outside the science content areas.

Notes

Structure and System Analyzers

CHAPTER 6

Picturing and Describing a System

Analyzers	Application
System Analyzer Structure Analyzer	· Characterize the composition of a system or structure in terms of its properties, elements, rules, and background space

Exploring the World with the Language of Patterns

When we look at the world we see structures and arrangements of structures into systems. We can divide the kinds of arrangements in the world into two major categories: those that are obvious on observation and those that are not. This is an important example of the progression of inquiry. Critical scientific analysis is used to understand both types.

    1. **Observed systems.** These are systems in which the elements or parts are arranged in a fashion that is obvious on observation. When we look at these systems, we try to understand how they are assembled and how they work. Typical examples are:

- Fields of naturalist study, such as:
  - Patterns of day and night,
  - Patterns in clouds and weather,
  - Patterns and behaviors of biological organisms,
  - Patterns of geological formation.
- Technical structures, such as:
  - Parts and schematics of machinery and tools,
  - Organizational charts of social groups (businesses, governments, law).
- Structures in artistic works, such as:
  - Chapters in novels,
  - Scenes and acts in dramatic works,
  - Stanzas in poetry,
  - Verses and refrains in songs,
  - Movements in instrumental works.

2. **Inferred systems.** In many systems, a group of objects are linked in a fashion that is not so obvious with simpple observation. The connections between the elements must often be inferred or deduced. Examples of these include:

- The classification of the chemical elements based on observable properties into the Periodic Table. The organization of the Periodic Table infers the internal, unobservable structure of atoms.
- The classification of living things into a systematic organization. This organization is necessary for the process of evolution to be inferred and understood.
- Developing an understanding of the motion and evolution of stars and planets from observations of the light and motions of the planets and stars in the sky. These explorations allow understanding of the birth and evolution of the expanding universe by inference.

- Understanding processes outside of the realm of "natural sciences". Examples include:
  - The hidden relationships between characters in a story or work of art.
  - The explorations of shape and form and how they relate to natural objects in the art of Picasso, Monet, or Leonardo da Vinci.
  - Bach's exploration of the ordering of notes in the fugue.

The first of these systems can be explored successfully with observation, while the latter requires the tools of logic and inference in addition to observation. Both require that we can explore a system or systems of objects and describe the elements, rules, and background space. Thus, all critical understanding starts with a careful Language of Patterns analysis.

The Graphical Analyzer System in Scientific Inquiry	Scientific investigation is concerned with inquiry into how a system is arranged. Sorting and grouping are natural tasks for the brain; in scientific investigations, they are the starting point for deeper inquiry into a system and its properties. This task is approached through a more thorough analysis of the properties and arrangement of the elements in the system.

The role of the Graphical Analyzer System is to guide the description of a system or structure for critical analysis. To appreciate the value of such a method, try this apparently simple (but deceivingly difficult) task:

In words, write a description of this system (structure) in the box. You will be surprised how hard it is to communicate the information clearly and completely.

Frustration is commonly felt from not knowing how to begin this task. This is the same problem shared by teachers and students at every grade level when confronted by word problems, essay questions, and other forms of open-ended questions.

- Read your description to a friend and have them draw it from your description.

- Compare your friend's drawing to the actual system.

- You will likely confirm the difficulty that you experienced in the description and communication of such an apparently simple system.

System Analyzer	Any system or structure or sub-system (element) can be described using a System Analyzer.

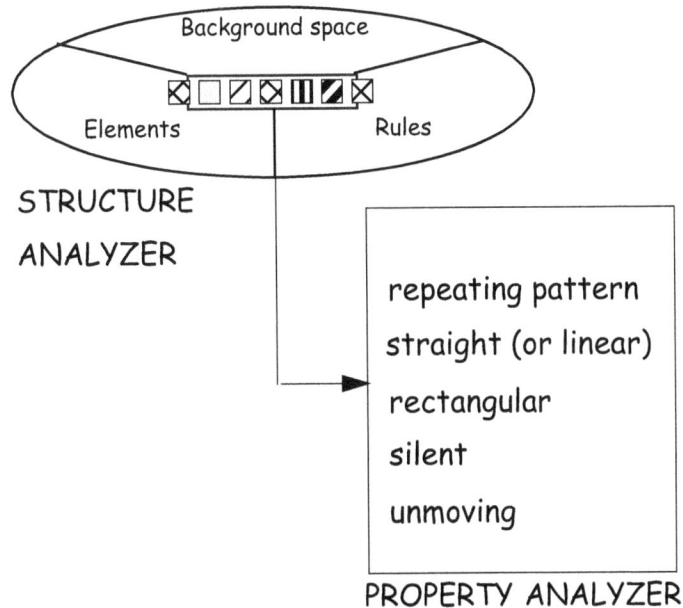

Background space

Elements

Rules

STRUCTURE

ANALYZER

repeating pattern

straight (or linear)

rectangular

silent

unmoving

PROPERTY ANALYZER

The System Analyzer has many of the characteristics of a simple concept map or "brain-storming" diagram currently in common use in many classrooms and work environments.

Structure Analyzer	The Graphical Analyzer System provides a methodical process to accurately describe and communicate that is widely required in science. The Structure Analyzer provides an overview of what we need to discover about a system in order to describe it scientifically. A language-of-patterns analysis forms the basis of describing structures:

- Every structure is comprised of a certain group of **elements**.

- In a structure, the elements are arranged somewhere - in what is called a **background space**.

· The **arrangement** of elements is governed by **rules of relationship** between each element and between the elements and the background space.

· The Structure Analyzer is differentiated from the system analyzer because the property analyzer has been removed. This emphasizes: **properties are emergent** from the interactions of elements, rules and background space.

This concept of structure is represented in the Graphical Analyzer System with a Structure Analyzer.

The Structure Analyzer can be used to quickly see the overall arrangement of an object or system of interest. A quick structure analysis often leads to a knowledge of what is needed to complete the description of the system.

Background space

B

Elements

Rules

A

C

Background space
B

A   Structure   C

Elements

Rules

STRUCTURE ANALYZER

This analyzer reminds us that structures are composed of a set of **elements** arranged by certain **rules** in a **background space** that will give rise to certain

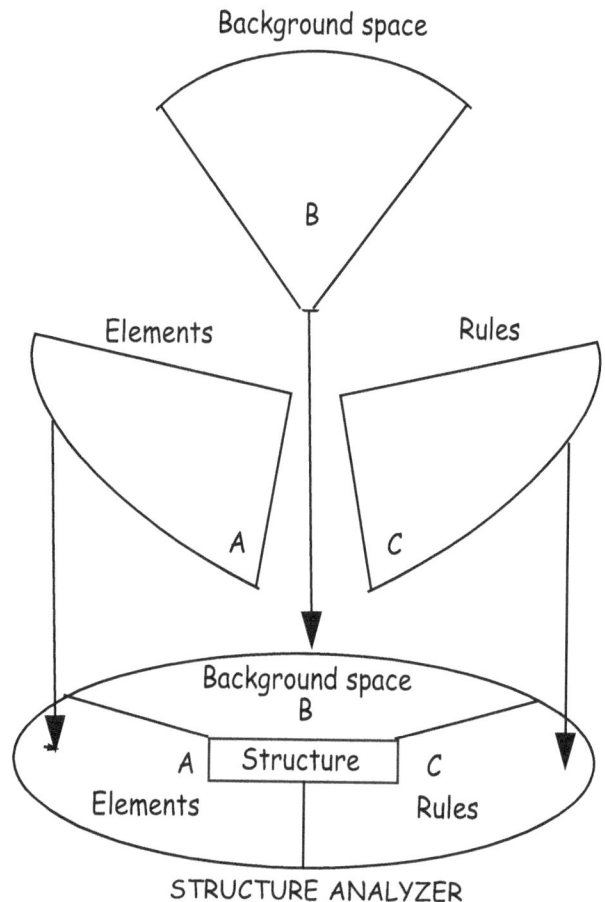

properties. The Structure Analyzer requires that its elements, rules, and background be defined and described.

In the pattern-filled squares structure example:

- The **elements** of this structure are the seven pattern-filled squares.

- The **background space** in which they are arranged is on a page of two dimensions (length and width).

- One of the **rules** that arranges the squares is that they are ordered in a straight line.

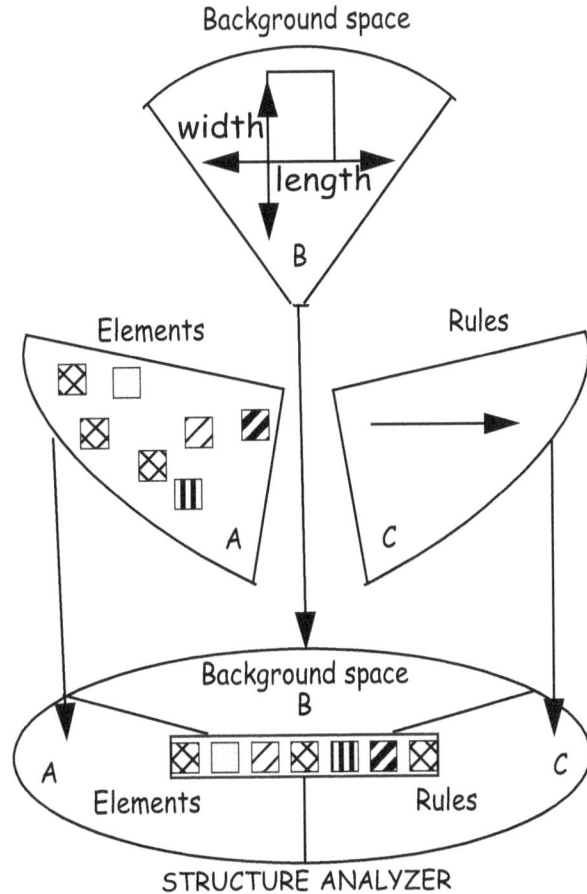

To fully describe the structure, rules to order the squares and to describe the internal fill pattern must also be included. This will be set forth in the coming chapters. In the next chapters, we will also explore the Graphical Analyzers that will help guide the discovery of elements, background space, and rules. Finally, we will see how to describe a system or structure when it changes or evolves.

Notes

CHAPTER 7

The Element Analyzer Series

Properties, Observables and Elements

Analyzers	Application
Property Analyzer	• Identify and characterize properties of objects and systems
Element Analyzer	
Element Counter	
Element Extractor	
Technical Writing Analyzer	• Systematically describe a pattern, structure, or system
Coding Grid	• Exchange objects and symbolic forms to build codes

Systems have **properties** that result from the arrangements of the elements in their background space. These properties are what give a system of structure its qualities that let us sort, group, and categorize it. Properties are not what makes up the structure or the system; they are the overall emergent <u>qualities</u> of the system. The properties that we measure and observe that describe the **state** of a system are called **observables**.

The System Analyzer

For reference, we repeat this graphic analyzer. A System Analyzer fully describes any system by adding the property analyzer to the structure analyzer.

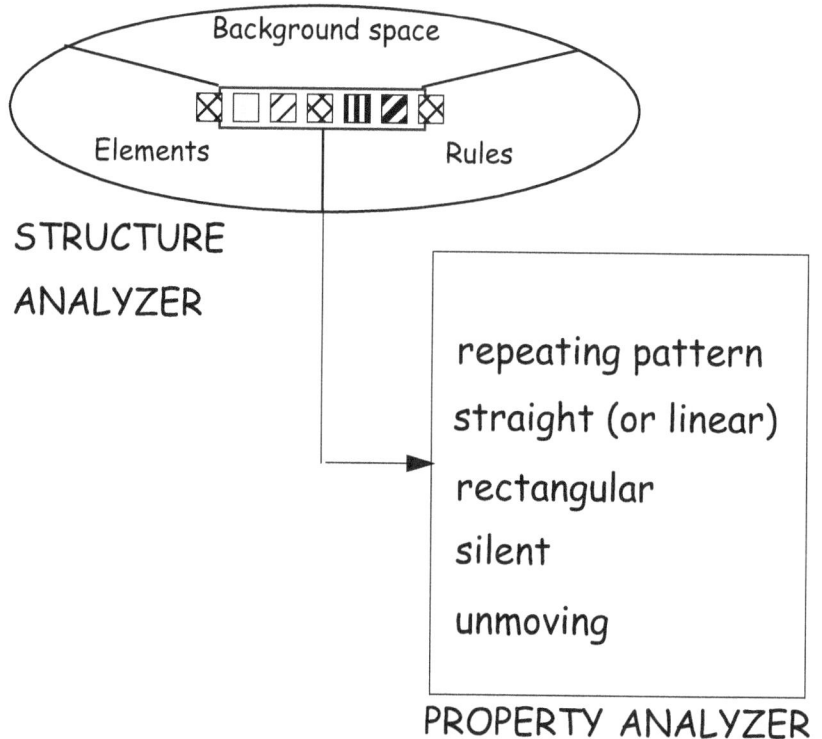

Background space

Elements   Rules

STRUCTURE

ANALYZER

repeating pattern

straight (or linear)

rectangular

silent

unmoving

PROPERTY ANALYZER

Any system or its parts [properties, elements, rules, background, changes] can be described using a **Property Analyzer**. In addition, it is can be useful to analyze a property as a system itself. This will focus analysis on how the emergent-property can be characterized and therefore measured [by choosing useful observables].

Property Analyzer

Some properties for the pattern system example have been entered on a Property Analyzer. Can you think of some others?

Properties do not have to be described only in words in these analyzers; examples of how these analyzers can be used with manipulatives and graphics are included in accompanying student tutorials and workbooks.

On the following page, a variety of named systems have been placed in the header section of a Property Analyzer. List some properties of the systems that are named. Properties are attributes and qualities of a system such as its color, texture, degree of difficulty, etc. Some properties have been filled in to get you started. These Analyzers may be arranged so that the properties used to describe a system become new systems that themselves can be analyzed. This process is also expressed in the more familiar, but somewhat less formal "concept map".

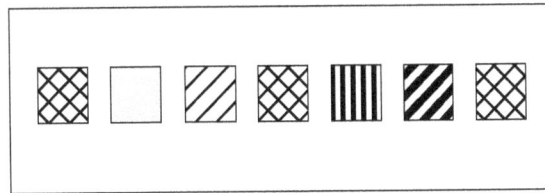

repeating pattern _ _ _
straight (or linear) _ _ _
rectangular _ _ _ _ _
silent _ _ _ _ _ _
unmoving _ _ _ _

PROPERTY ANALYZER

Practice Exercise

Complete the following exercises using the Property Analyzer. (Possible answers appear on the next page.)

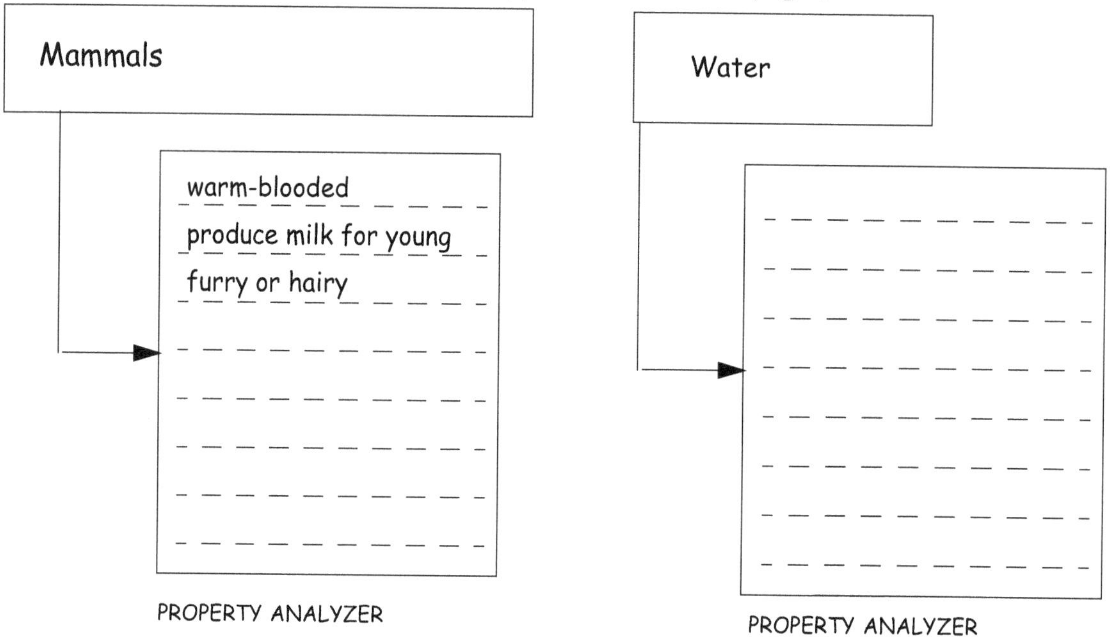

Mammals

warm-blooded
produce milk for young
furry or hairy
___ ___ ___ ___ ___
___ ___ ___ ___ ___
___ ___ ___ ___ ___
___ ___ ___ ___ ___

PROPERTY ANALYZER

Water

___ ___ ___ ___ ___ ___ ___
___ ___ ___ ___ ___ ___ ___
___ ___ ___ ___ ___ ___ ___
___ ___ ___ ___ ___ ___ ___
___ ___ ___ ___ ___ ___ ___
___ ___ ___ ___ ___ ___ ___
___ ___ ___ ___ ___ ___ ___
___ ___ ___ ___ ___ ___ ___

PROPERTY ANALYZER

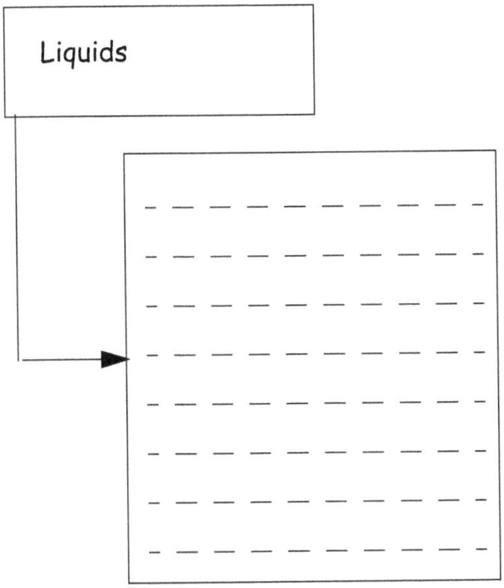

Liquids

PROPERTY ANALYZER

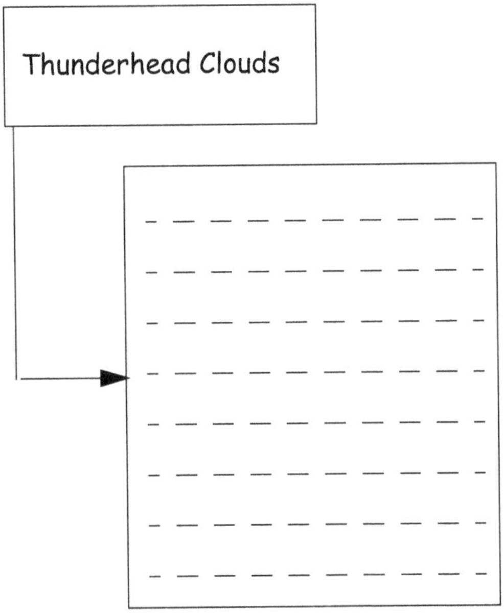

Thunderhead Clouds

PROPERTY ANALYZER

Answer   Possible answers to the exercises on the previous page.

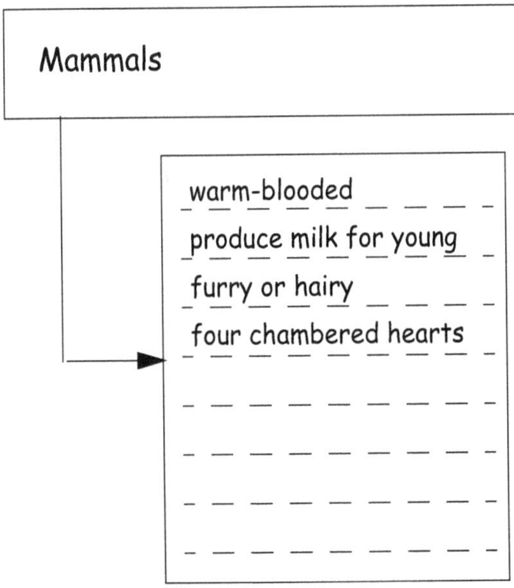

Mammals

warm-blooded
produce milk for young
furry or hairy
four chambered hearts

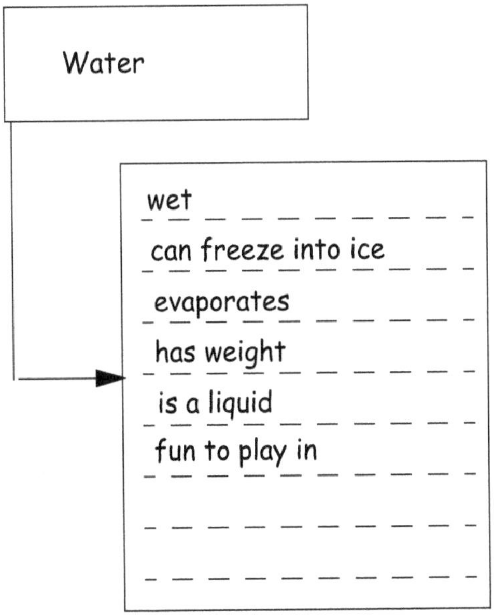

Water

wet
can freeze into ice
evaporates
has weight
is a liquid
fun to play in

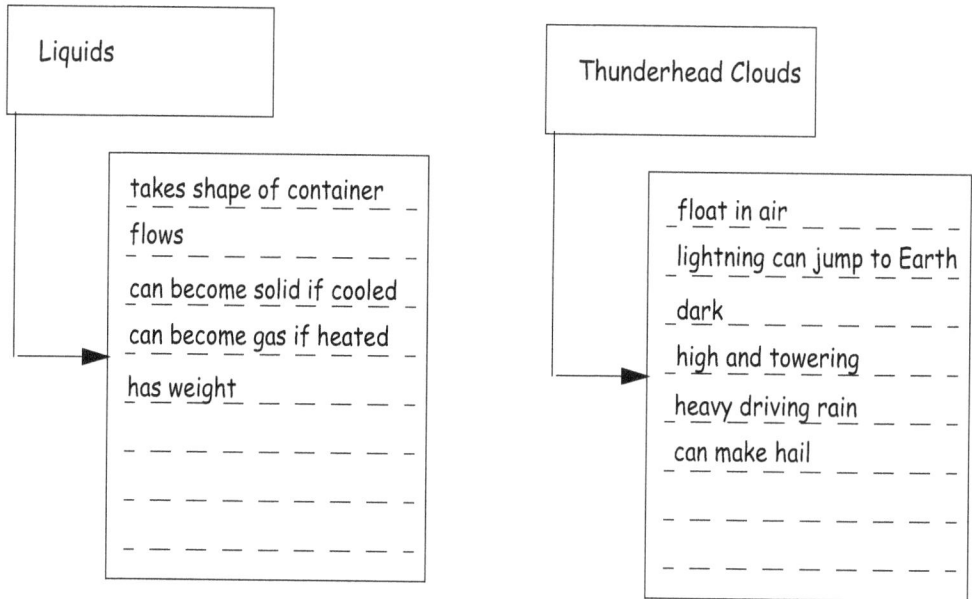

Liquids

takes shape of container
flows
can become solid if cooled
can become gas if heated
has weight

Thunderhead Clouds

float in air
lightning can jump to Earth
dark
high and towering
heavy driving rain
can make hail

Cross-
Curricular
Example

Here are several examples using the Property Analyzer across topic domains.

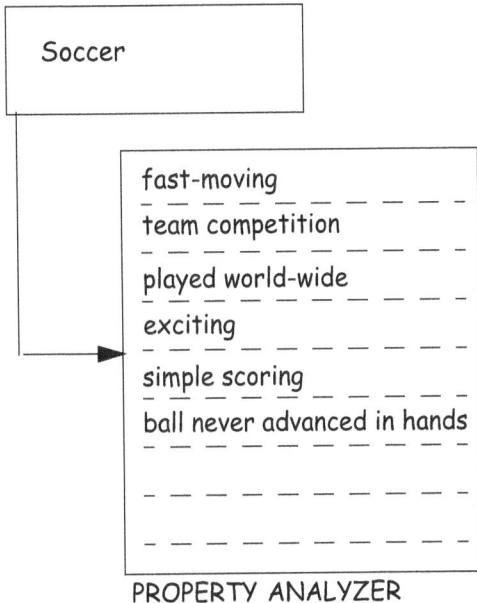

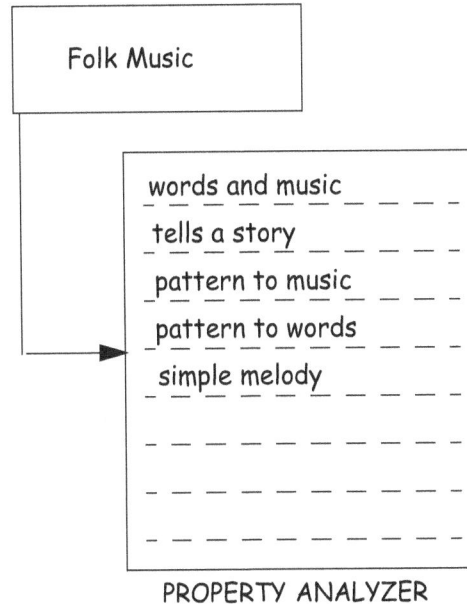

Soccer

fast-moving
team competition
played world-wide
exciting
simple scoring
ball never advanced in hands

PROPERTY ANALYZER

Folk Music

words and music
tells a story
pattern to music
pattern to words
simple melody

PROPERTY ANALYZER

**Element
Analyzer**

We can explore the example of the geometric pattern using the Element Analyzer.

If the elements in the pattern example are sorted by the **observable** of their fill pattern, the system of squares can be broken down. This manipulation separates the elements of the system into its individual parts. The categorization of the elements by observables suggests that they are "sub-systems".

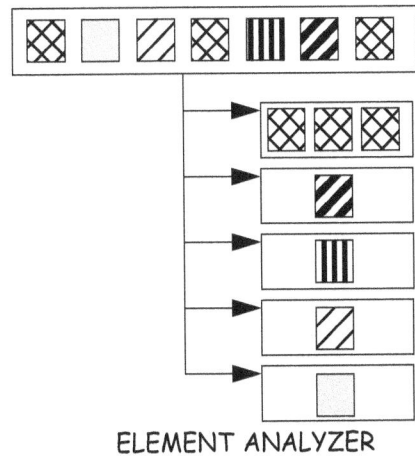

ELEMENT ANALYZER

Identifying the elements of an observation is an important scientific activity. A first grade sky watch might generate the following Element Analyzer.

Do Not Underestimate the value of this as a first step in discovery at every level of knowledge. When you encounter something you have never seen before, start by listing the elements of system.

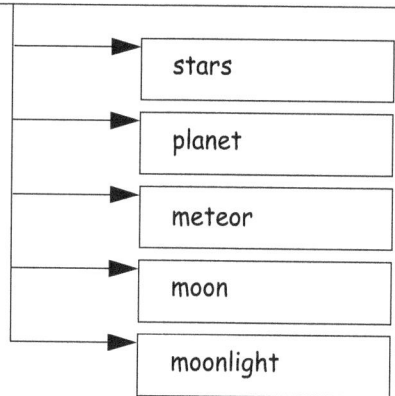

stars

planet

meteor

moon

moonlight

The accessibility of this "simple" activity to early learners should not diminish its importance at very sophisticated levels in science. This analysis is widely used from medicine to particle physics. For example, it can be used to diagnose the presence of a genetic disease by analyzing the pattern of blood proteins by electrophoresis in a medical laboratory (right).

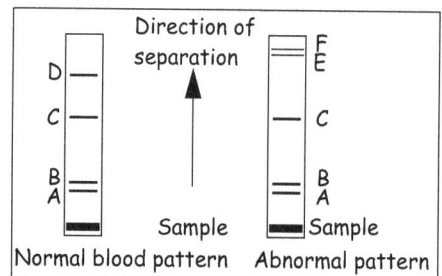

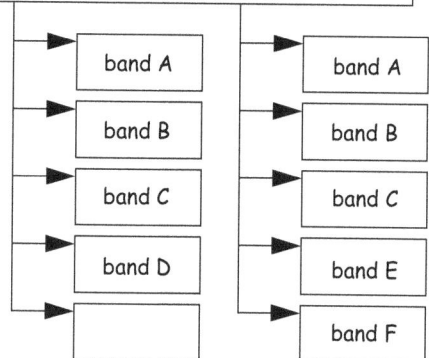

Direction of separation

Normal blood pattern   Abnormal pattern

band A		band A
band B		band B
band C		band C
band D		band E
		band F

Element Counter

The basic Element Analyzer can be modified to allow the separated elements to be described in more detail.

The first modification is to count the number of elements in the group. This is the **Element Counter**:

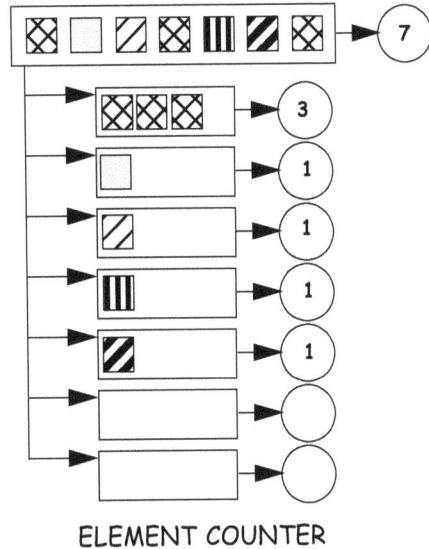

ELEMENT COUNTER

Here is another example of counting the elements of a system with the **Element Counter**:

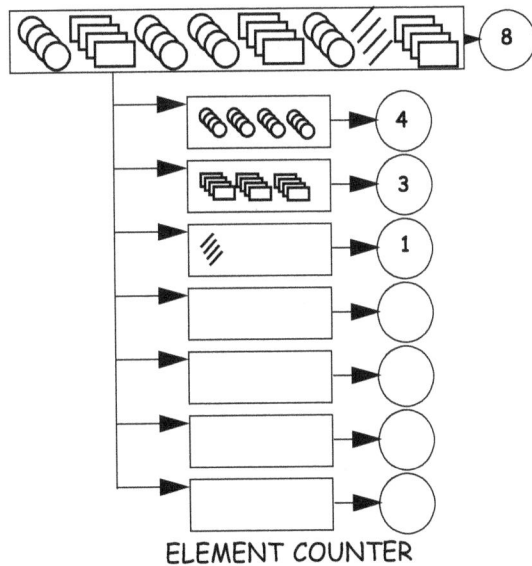

ELEMENT COUNTER

The Element Extractor

We have already learned that each element in a system has its own structure, i.e., it is a sub-system. The Element Analyzer can be extended to further define the subsystems or substructures in a system. This is the **Element Extractor**.

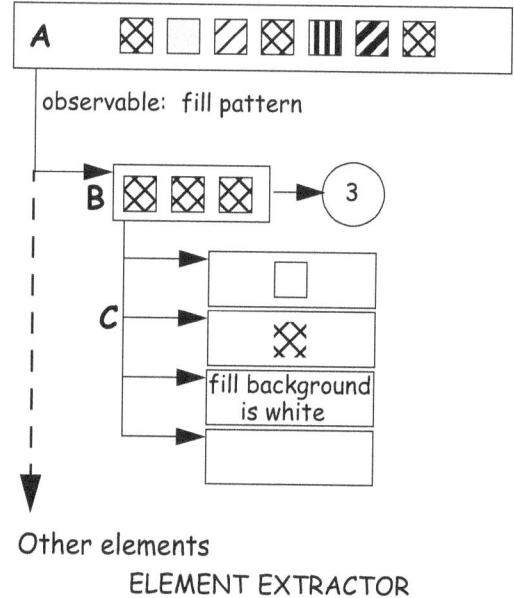

observable: fill pattern

Other elements

ELEMENT EXTRACTOR

In this example, a single element is being further analyzed. The Element Extractor has extracted three similar squares (**B**) from the overall group of seven (**A**) [the observable - fill pattern - is listed]. The properties of these similar squares are further extracted in the third tier of the analyzer (**C**).

In the fourth tier (**D**), the cross-hatch pattern is a system itself and can be further extracted.

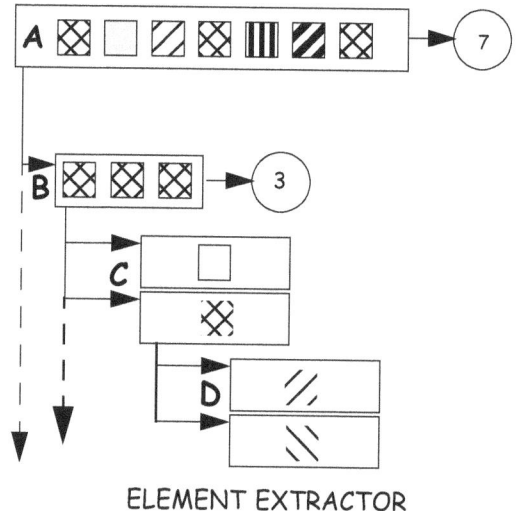

ELEMENT EXTRACTOR

The Elements of Systems Modeling

The Element Analyzer: Single Attributes Practice Exercise

The elements in the systems that we have been looking at have had only two attributes: a uniform geometry and a varying fill pattern. Draw an Element Analyzer when the elements in a system share <u>every</u> attribute (i.e., each element is identical).

Answer

Your figure could resemble this one:

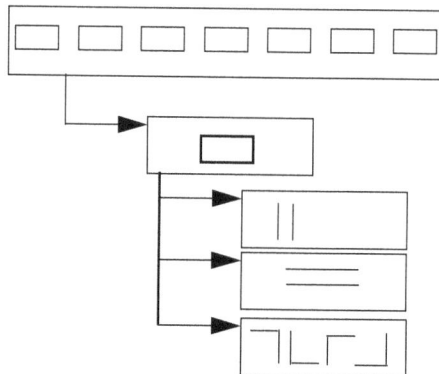

Below are two examples of a single attribute analysis in different situations. The first is a chemistry example, and the second is speech and drama.

	Structure A	Structure B
Chemistry	(–◯–)	(–◯–◯–◯–◯–◯–◯–◯–◯–◯–◯–◯–◯–)
Speech	"EE"	"EEEEEEEEEEEEEEEEEEEEEEEEEEEEEEEEEEEEE"

Each structure is s system made from a monotonous series of elements. In the chemistry examples, the chemical structure in **A** is related to swamp gas, and the chemical structure in **B** is related to animal fat. In the speech examples, the form in structure **A** is an interjection or comment, while the speech structure in **B** is a primal scream.

This analysis shows how the arrangement of the elements, even when they are the same, can lead to very different properties in systems and structures. This property of repeating elements is a key to understanding organic chemistry, biochemistry, and the repeating structure of crystals in nature.

Technical Writing Analyzer

All of the properties we have extracted so far can be labelled. If we choose to label them in a written language, the grid would look like this:

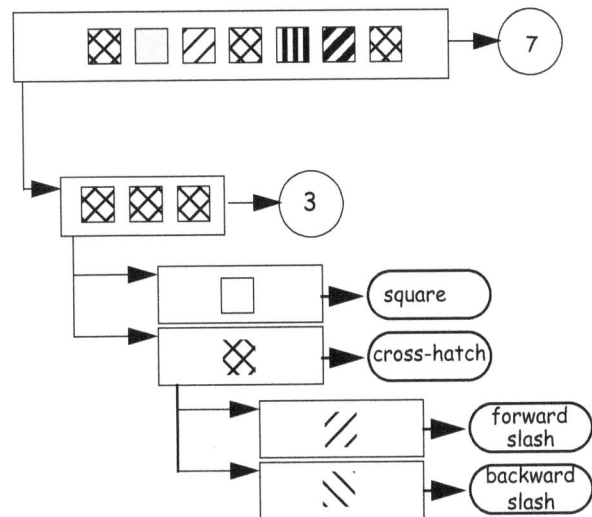

TECHNICAL WRITING ANALYZER

This extension of the Element Analyzer is called a **Technical Writing Analyzer**. The extraction of the information in this highly complex structure should not be underestimated. If this grid is now read in English, a technical summary of the extracted element can be easily and precisely composed:

> In this structure (system) there are seven squares.
> Three are filled by a cross-hatched pattern.
> The cross-hatched pattern is made from forward and
>   backward slashes.

The properties used in the Technical Writing Analyzer do not have to be expressed in English.

Technical Writing Assistant

Spoken and written language is the first and most common symbolic language used by most people. However, the following example will demonstrate that this process of Technical Writing can be applied for precise analysis and description in any field of interest.

If the expression is a chemical structure, then chemical symbols will translate the structure into the formal language of chemical notation. Two such examples follow.

This first example uses the Technical Writing Analyzer to decode the cleaning fluid, carbon tetrachloride; it might be used in a middle school science class learning chemistry terminology.

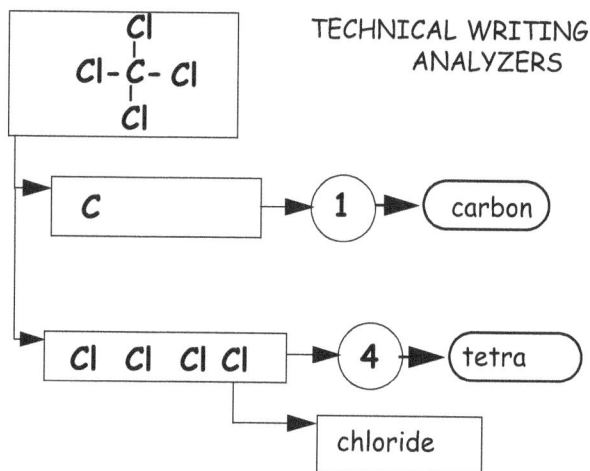

Translation: carbon tetrachloride (CCl_4)

This second example might be seen in a more advanced, University level organic chemistry class. Both use the same process that has already been described

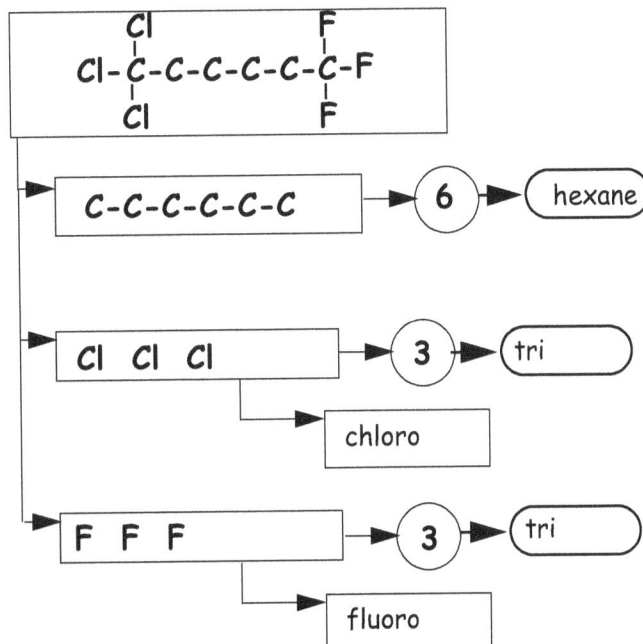

Translation: trifluoro, trichloro-hexane

This is a very useful tool. This process takes an observation through a cycle of critical analysis to provide an expression of the observation in a formal language. This is a process of **formal modeling**. This type of formal modeling is useful for communication of ideas and descriptive data, a vital initial step in the scientific method of inquiry.

Cross-
Curricular
Example

To show the breadth and power of the Technical Writing Graphical Analyzer, a football play is diagrammed. This is a method of annotating the complex movements of 11 players. When a play is called in the football huddle, a cryptic language is used that provides all of the information needed to put every player in place and then into action. In this case, the play is "T-Right-34-Trap". This notation arranges the linemen and backfield players, assigns blocking patterns, and details ball handling instructions to create an opening in the defensive positions. The fullback (the "3" back) is given the ball and runs into this opening. The players execute the play physically after receiving the verbal description.

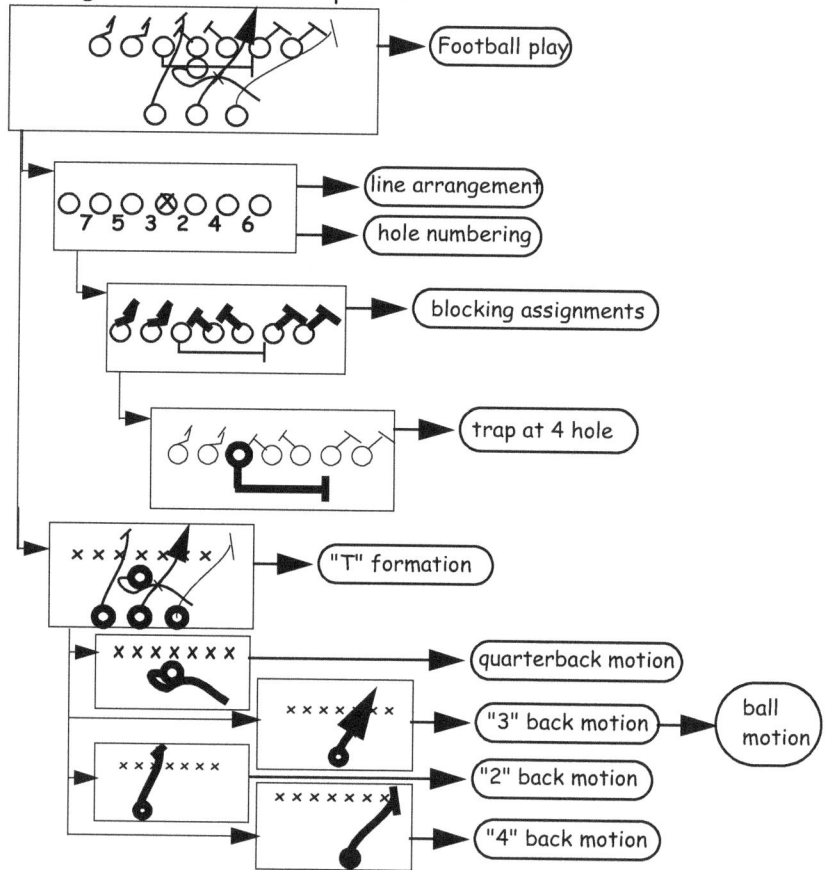

Football play

line arrangement

hole numbering

blocking assignments

trap at 4 hole

"T" formation

quarterback motion

"3" back motion

ball motion

"2" back motion

"4" back motion

TRANSLATION: T-Right-34 Trap

Coding Grid

A Coding Grid guides the re-assignment of an element in one system into a symbolic element in a second system. Coding Grids are more directed than the closely related technical writing analyzers that extract and summarize information.

A simple one-to-one attribute exchange can be performed by students as early as first or second grade. More sophisticated translation and coding can be expected as age and education progresses. The Coding Grid guides an essential modeling process of mapping an observation into a symbolic system using the following steps:

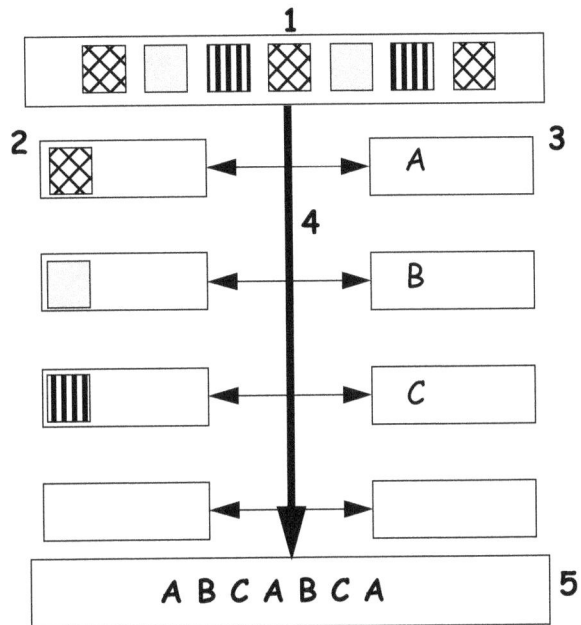

CODING GRID

1. The appropriate observables are chosen from the **system** to be modeled.

2. The system is described using the elements, arrangement, rules, and background space.

3. A corresponding set of **symbols** is selected.

4. The observable properties are mapped onto the symbolic forms.

5. The symbols can then be used to represent or **model** the original system.

Practice
Exercise

Students experimenting with batteries, switches, and bulbs to make circuits can use a Coding Grid to report their work using standard electronic symbols. Complete this Coding Grid to draw a correct schematic for the circuit shown.

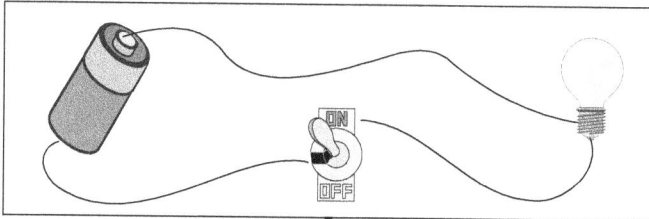

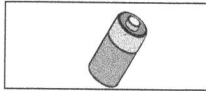

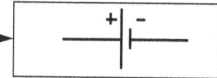

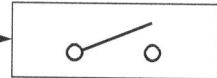

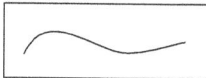

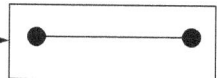

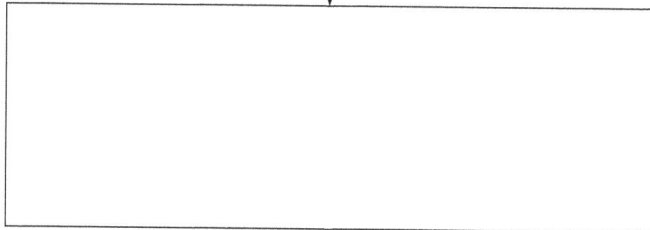

- The answer is given below:

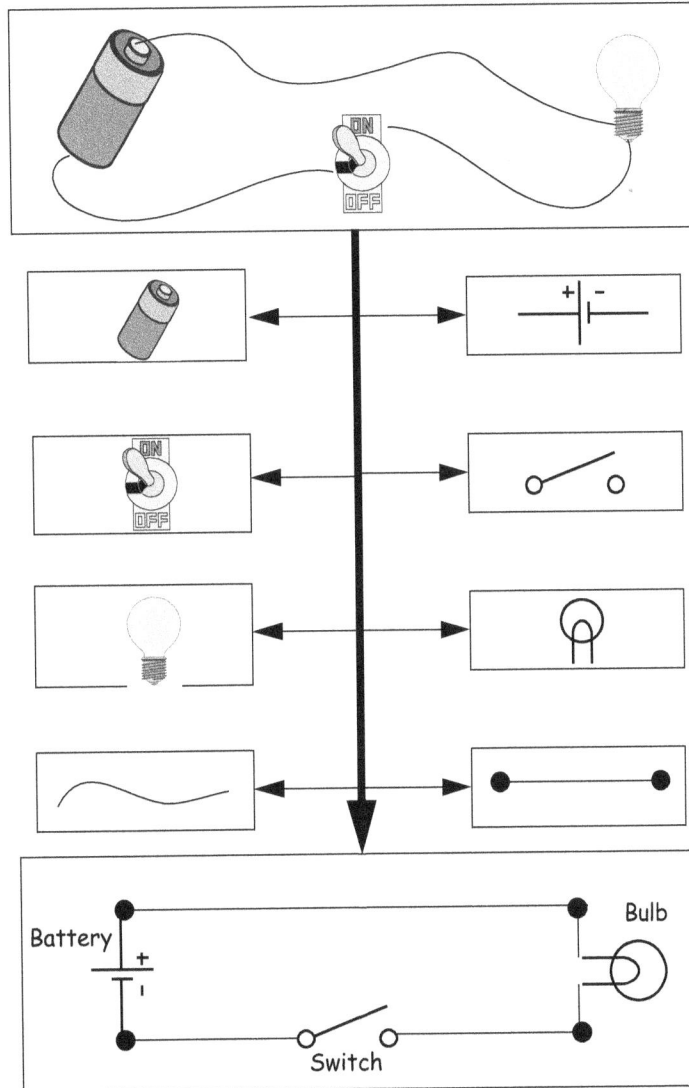

Notes

CHAPTER 8

Sorting and Classifying with the Graphical Analyzer System

Observables and Elements

Analyzers	Application
Sorting Analyzer	• Sort, group, classify, and sub-sort
Multiple Observable Sorting Analyzer	
Classifying Analyzer	• Make classifying keys and perform logical sorting

Sorting and Grouping

When any system or structure of interest is observed, the natural first step is to group and sort the various elements. The grouping and sorting is always based on some attribute of the elements that is chosen by the observer. Often this attribute may not be clearly stated. Therefore, in critical thinking, the first step is to recognize what attribute is being used or observed in order to begin the sorting and grouping process. The chosen attribute is the **observable** used to perform the sort. The Graphic Analyzers presented in this chapter are useful for sorting, classifying, and exploring how observables and attributes affect a classification scheme.

Often grouping and sorting by "obvious" observables does not lead to a complete understanding of a system. In these cases, the scientific approach is to make a more complete exploration of the full set of properties in the system. In the next chapter, we will tour the Graphical Analyzers used for these more extensive investigations.

Sorting Analyzer: Single-Sort

A system of objects can always be partly described in terms of its elements. Here, a Sorting Analyzer will be used to categorize elements in a system.

The oval space in the center of the Analyzer states the observable. The observable is the property that has been chosen by the observer to evaluate the elements. Here the observable chosen is "shape". The sort that results from choosing this observable is shown. In this instance, shape is probably not a useful choice to describe this system.

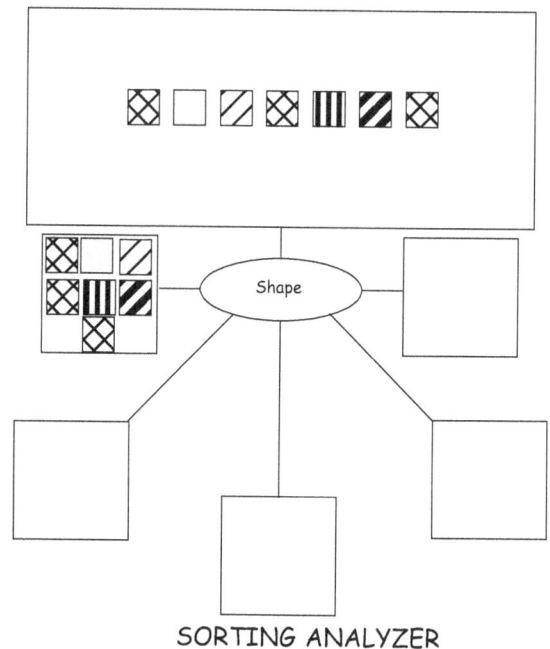

SORTING ANALYZER

Perhaps a better choice of observable would be having "lines inside". The observable of having "lines inside" effectively selects the stippled block from the others. This is a practical result, but its usefulness depends on the questions being asked by the observer.

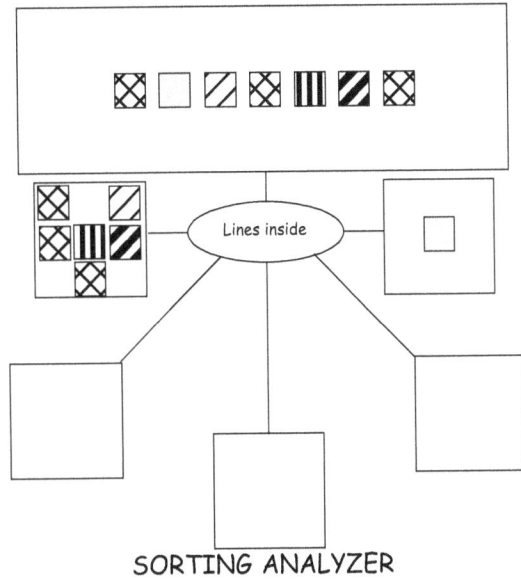

SORTING ANALYZER

If we want to know how many different kinds of blocks there are, none of the previous observables is useful. Instead we would choose "fill pattern" as the observable.

Using fill-pattern as the observable, the Sorting Analyzer sorts the square pattern system similar to that of the Element Analyzer.

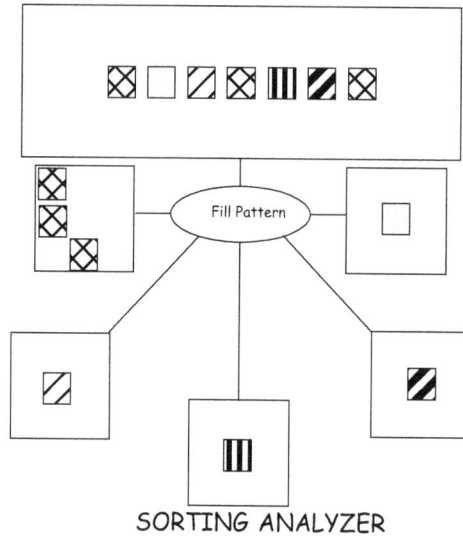

SORTING ANALYZER

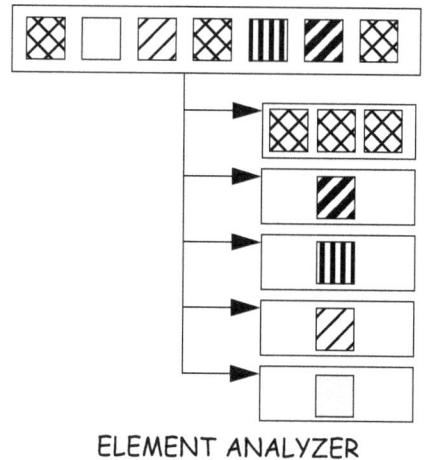

ELEMENT ANALYZER

Practice
Exercise

Sort these animals using the labeled observables. (The animals can be identified by name or given a letter symbol to sort.) Animals with the same characteristic of the observable should be grouped together. For example, in the first sort (number of legs), the sorting bins would contain animals with no legs, 1 leg, 2 legs, 4 legs, etc. Possible answers on next page.

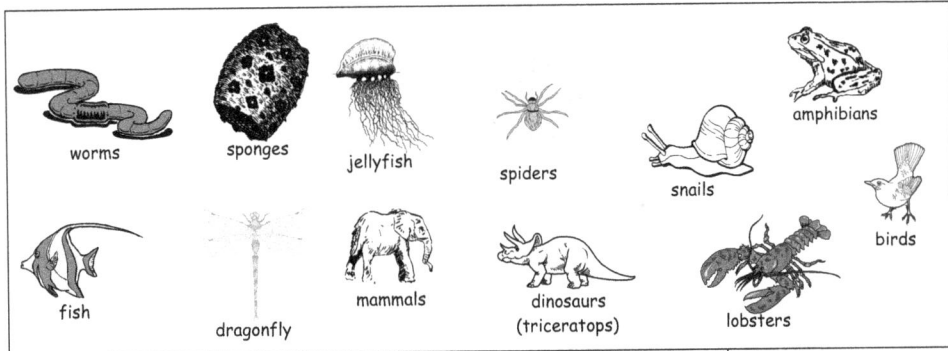

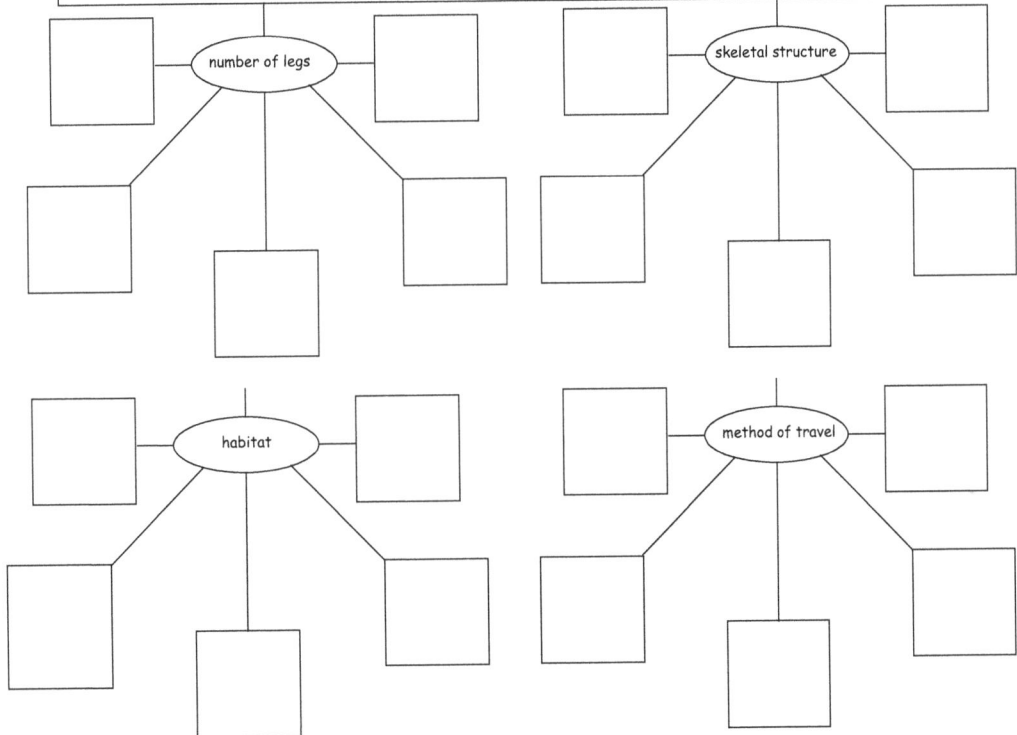

worms   sponges   jellyfish   spiders   snails   amphibians

fish   dragonfly   mammals   dinosaurs (triceratops)   lobsters   birds

number of legs

skeletal structure

habitat

method of travel

The Elements of Systems Modeling

77

Answer Sorting Analyzer: Poly-Sort

This is an answer to the animal sorts. If different observables sort the same group at the same time, a poly(morphic) sorter is made. This answer has been drawn as a polymorphic sort.

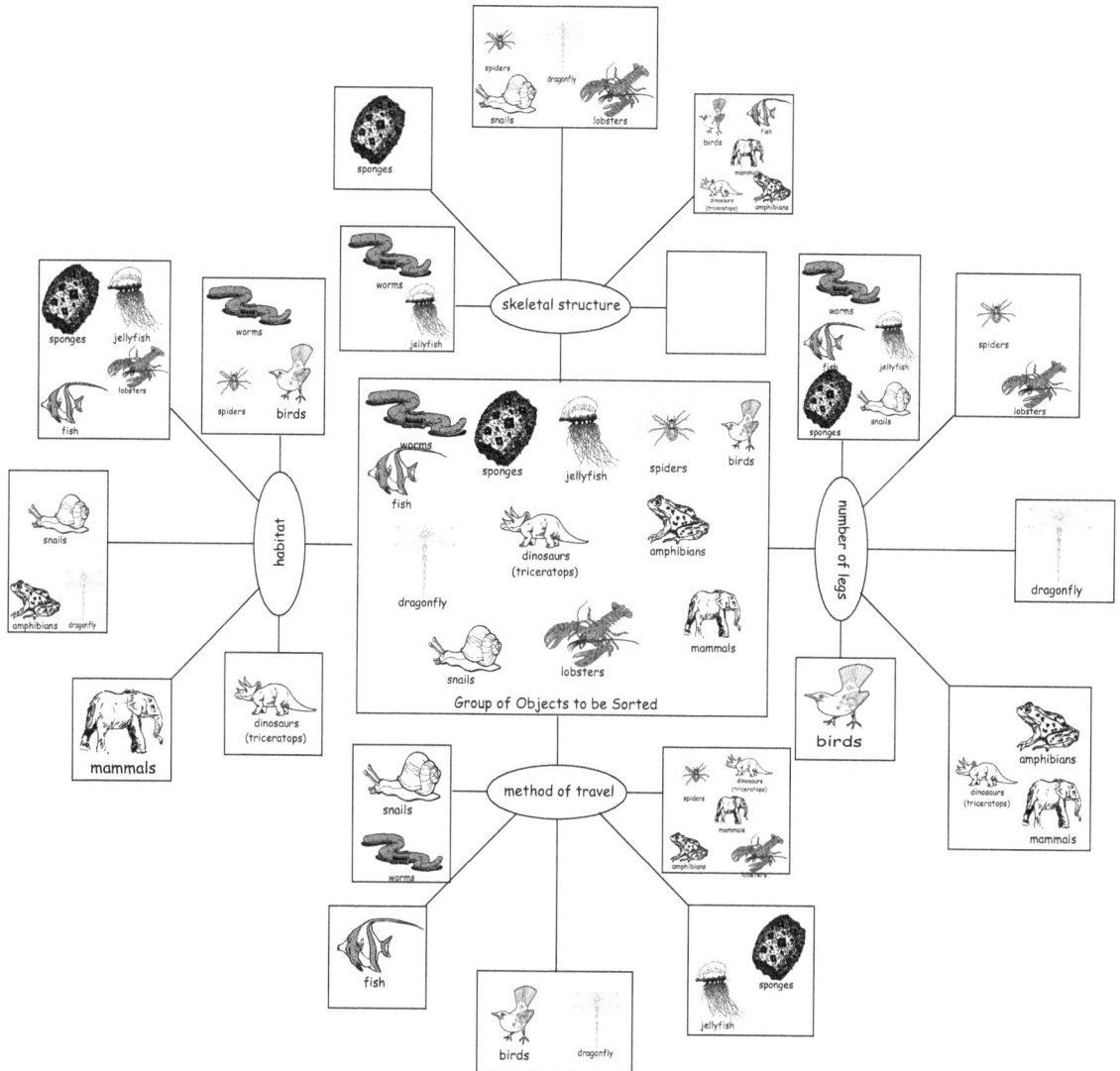

POLYMORPHIC SORTING ANALYZER

The Sorting Analyzer: Multiple Attributes

The purpose of the Sorting Analyzer is to allow a methodical categorization of the elements of a system. Many systems have multiple attributes which can lead to a variety of sorting strategies. One strategy may be more useful for answering certain questions than another strategy. It is often useful to be able to sort the same group of objects by different observables. This is an important way to search for connections within a group. When searching for connections, each sort pattern can be compared in a Compare and Contrast Analyzer. Consider this system:

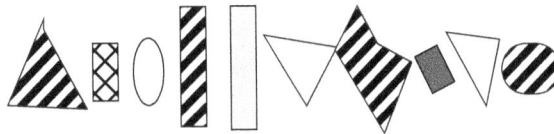

Practice Exercise

Try sorting this grouping by the different attributes or observables listed in the observable oval. Possible answers appear on the next page.

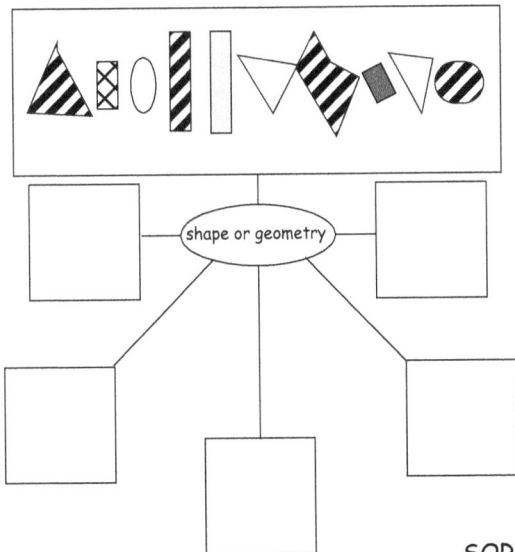

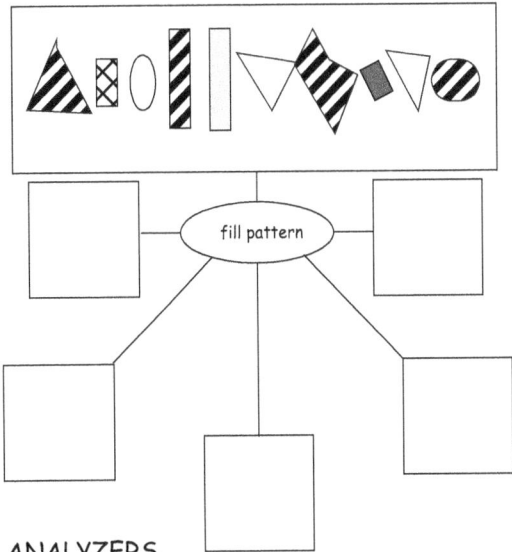

SORTING ANALYZERS

Answer

Here are some possible solutions for the previous sorts.

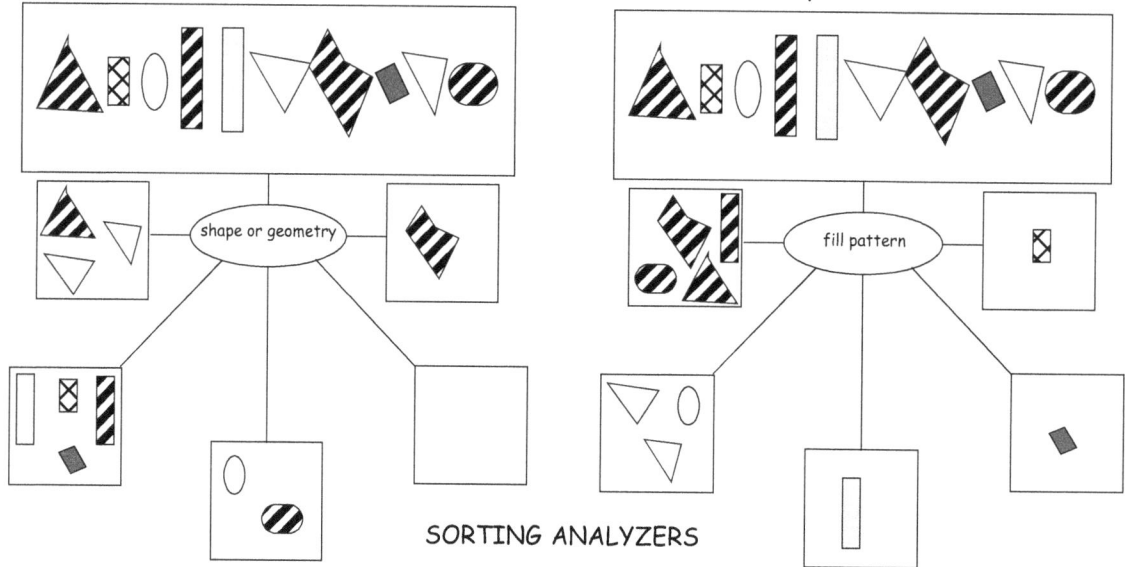

shape or geometry

fill pattern

SORTING ANALYZERS

Practice Exercise

By reversing the exercise, inference skills can be developed by asking, "What observable would have created this sort pattern?" This is not as easy as sorting by observable!

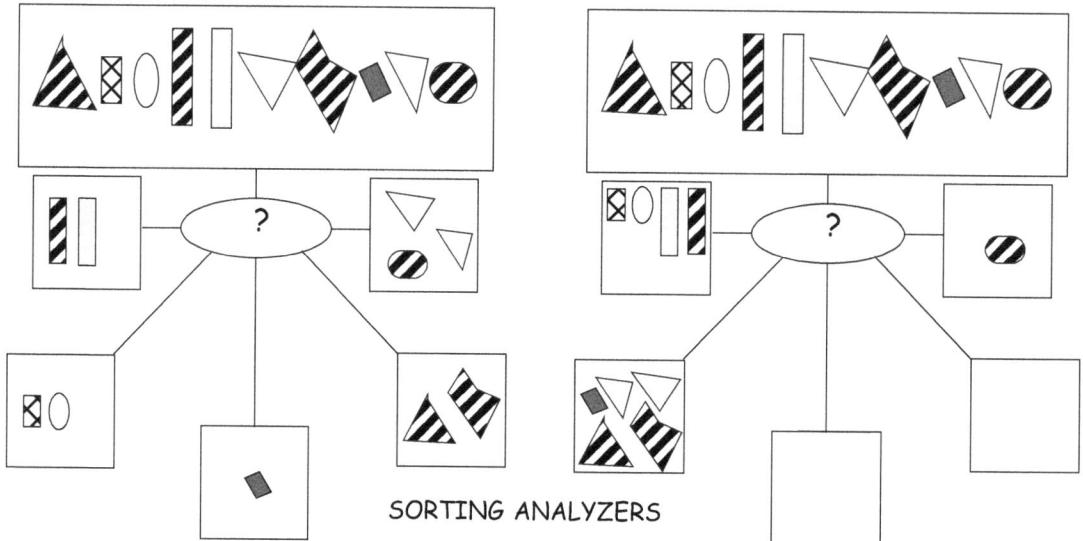

?

?

SORTING ANALYZERS

In the exercise on previous page, the observable in sorting analyzer on left is <u>size</u>; the observable in Sorting Analyzer on right is <u>long axis orientation.</u>

Sub-Sorting Analyzer

Finally, sub-sorts can be performed on previously sorted elements.

Practice Exercise

Try this example. One of the elements in the sub-sorting bin has been entered. [Answer on the following page.]

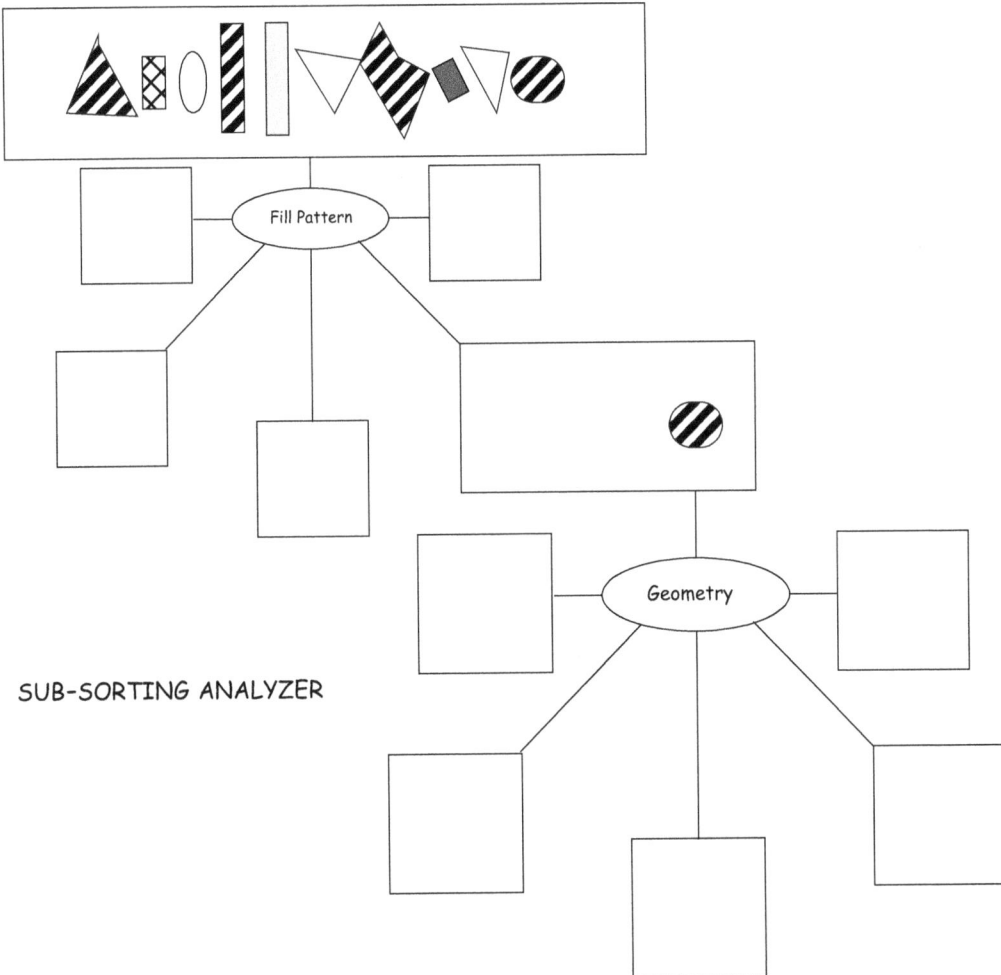

SUB-SORTING ANALYZER

Answer   This is a sequential, single observable sort and single observable sub-sort of a set with multiple attributes.

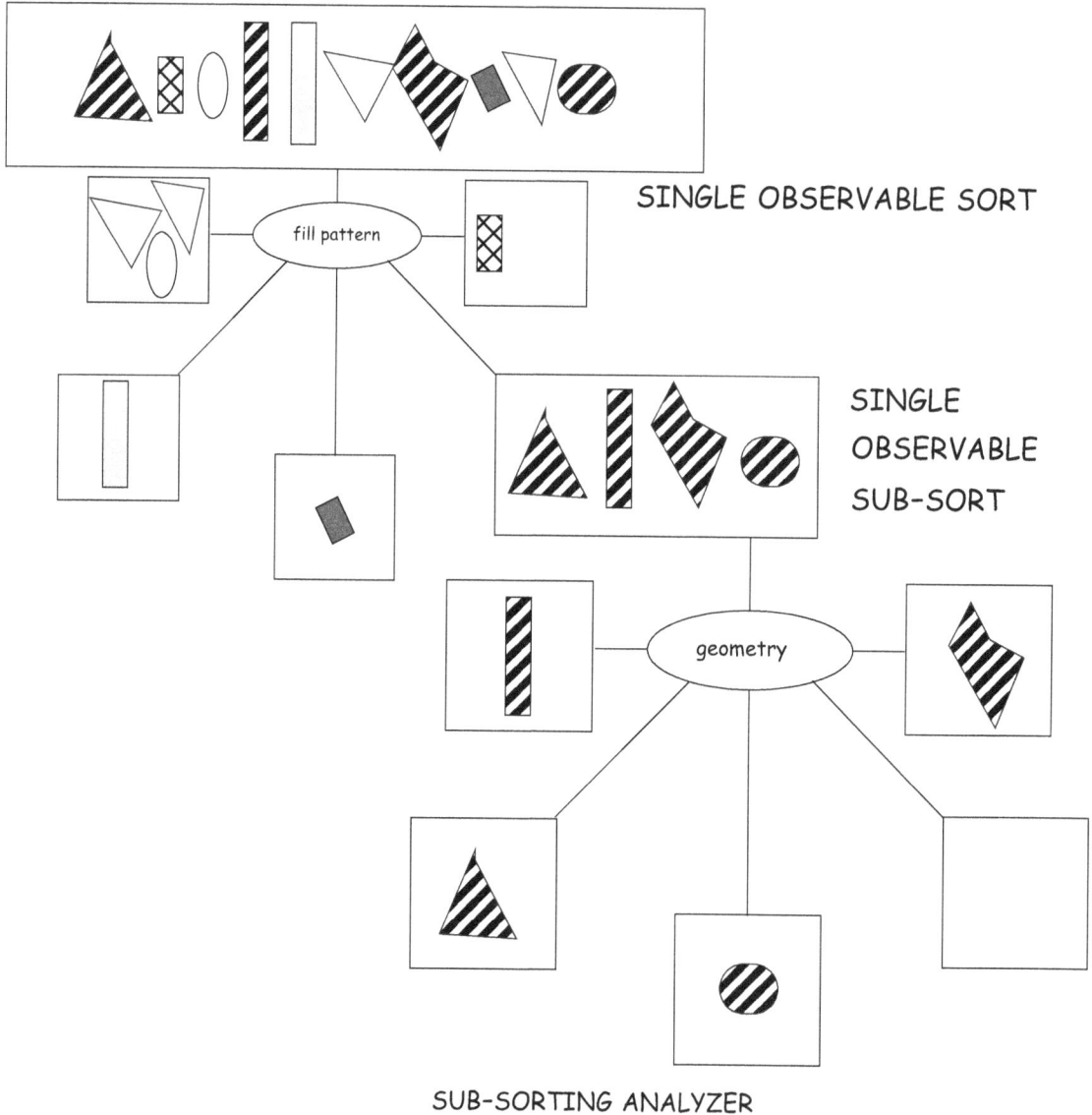

SINGLE OBSERVABLE SORT

SINGLE OBSERVABLE SUB-SORT

SUB-SORTING ANALYZER

Practice
Exercise

Sub-sorting is a very important classification process in scientific inquiry. In this example, chemical elements may be classified by physical state and by the types of chemical reactions that they undergo. First, sort this group of chemical materials by their state of matter at room temperature (25° C). Next sort the gases according to their pattern of chemical reaction. For many of us, this information will not be at our fingertips and may require further research into the science content of specific chemical reactions. For the student, this turns the question of process into a minds-on research project.

Chemical Identification

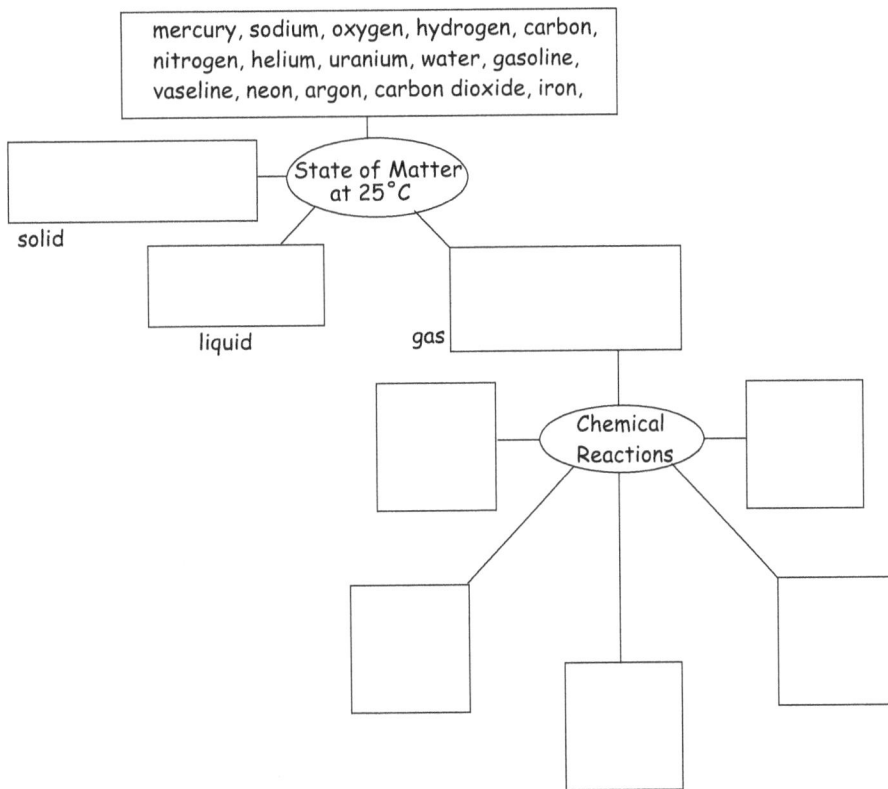

mercury, sodium, oxygen, hydrogen, carbon, nitrogen, helium, uranium, water, gasoline, vaseline, neon, argon, carbon dioxide, iron,

State of Matter at 25°C

solid

liquid

gas

Chemical Reactions

SUB-SORTING ANALYZER

Cross-Curricular Exercise

As in the previous example, sports and games are a natural area for sub-sorting practice. First, separate the list of recreational activities into the labelled game types, an activity may be used more than once at this level. Next separate the groups by the second attribute listed.

Rook, Monopoly, Crazy Eights, Hearts, Soccer, Football, Hide and Seek, Old Maid, Go-Fish, Slap, Bridge, Scrabble, Kickball, Baseball, Softball, Rubric's Cube, Tag, Candy-land, Chutes and Ladders, Hares and Hounds, Anagrams, Kick the Can

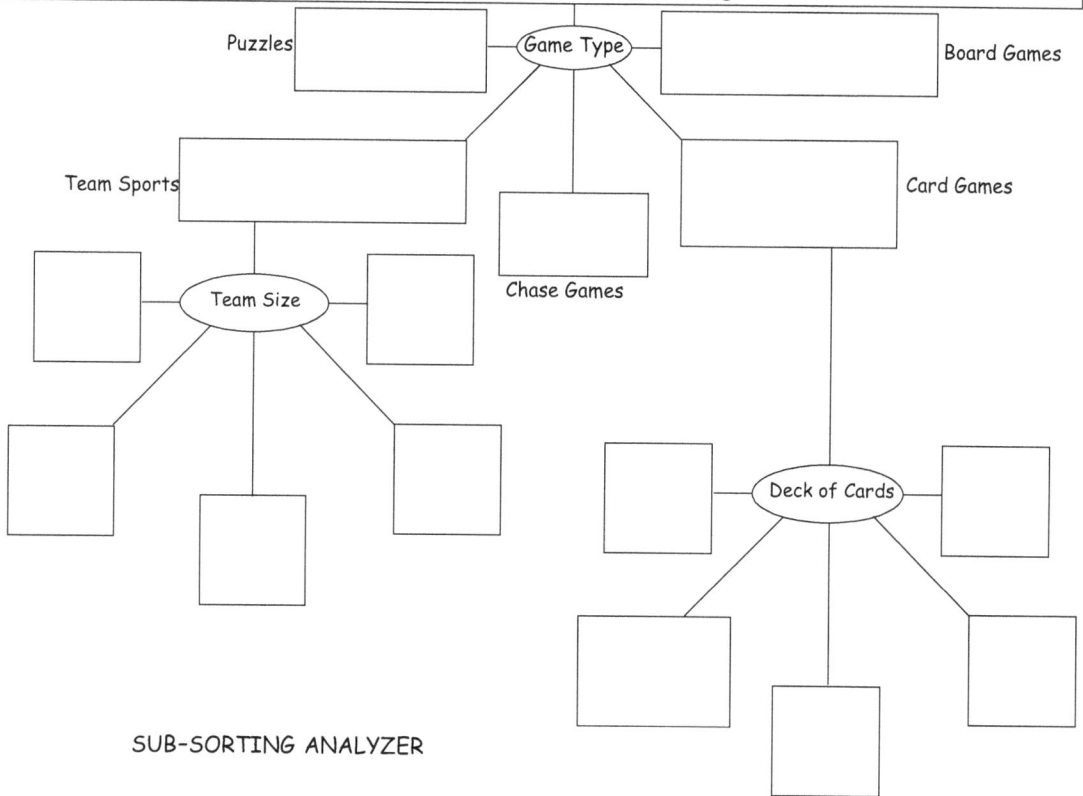

Puzzles   (Game Type)   Board Games

Team Sports   Card Games

Team Size

Chase Games

Deck of Cards

SUB-SORTING ANALYZER

Answer

Answers to the previous sub-sorts.

Chemical Identification

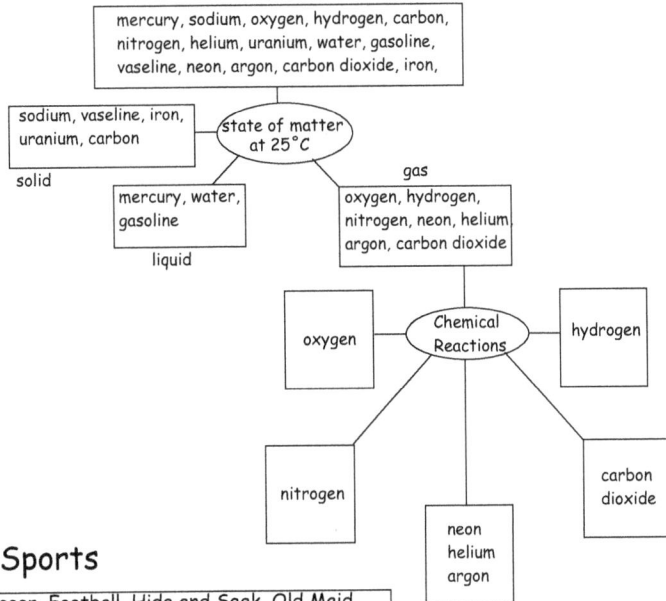

mercury, sodium, oxygen, hydrogen, carbon, nitrogen, helium, uranium, water, gasoline, vaseline, neon, argon, carbon dioxide, iron,

sodium, vaseline, iron, uranium, carbon

state of matter at 25°C

solid

gas

mercury, water, gasoline

liquid

oxygen, hydrogen, nitrogen, neon, helium, argon, carbon dioxide

oxygen

Chemical Reactions

hydrogen

nitrogen

carbon dioxide

neon
helium
argon

Games and Sports

Rook, Monopoly, Crazy Eights, Hearts, Soccer, Football, Hide and Seek, Old Maid, Go-Fish, Slap, Bridge, Scrabble, Kickball, Baseball, Softball, Rubric's Cube, Tag, Candy-land, Chutes and Ladders, Hares and Hounds, Anagrams, Kick the Can

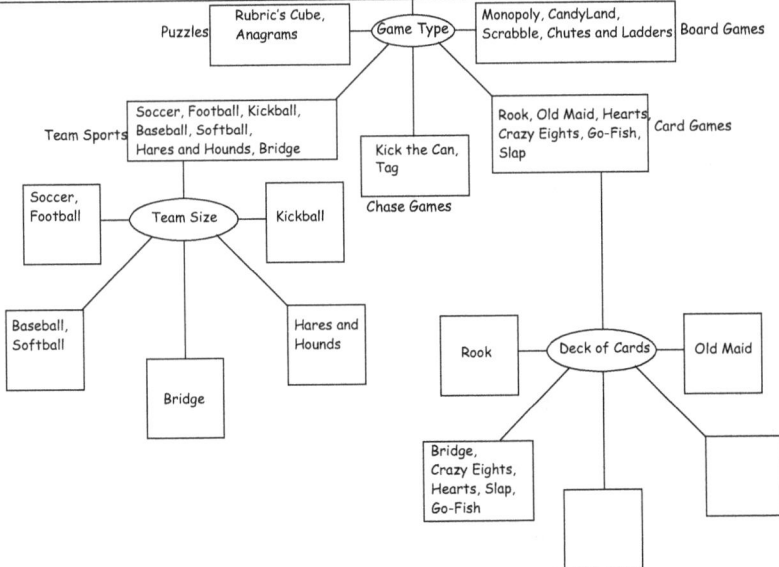

Puzzles

Rubric's Cube, Anagrams

Game Type

Monopoly, CandyLand, Scrabble, Chutes and Ladders

Board Games

Team Sports

Soccer, Football, Kickball, Baseball, Softball, Hares and Hounds, Bridge

Kick the Can, Tag

Rook, Old Maid, Hearts, Crazy Eights, Go-Fish, Slap

Card Games

Soccer, Football

Team Size

Kickball

Chase Games

Baseball, Softball

Hares and Hounds

Bridge

Rook

Deck of Cards

Old Maid

Bridge, Crazy Eights, Hearts, Slap, Go-Fish

Classifying Analyzers

The Sorting Analyzer can be modified to form a **Classifying Analyzer** [sometimes these are called dichotomous or binary keys].

The mental process of sorting and grouping tests objects for "sameness". If two things are the same according to the observable or property used in the comparison, then they can be grouped together.

The use of the Sorting Analyzer to sort from a general group has been demonstrated. The next step is to see how the Analyzer can be used to separate and classify objects out of a group. This analytical process is very important throughout scientific classification. It also has an important role in problem solving, logic, mathematics, law, and all forms of art criticism.

Here is how a Classifying Analyzer can be made:

1. First, the Sorting Analyzer is designed to separate the larger set into two groups.

SORTING ANALYZER

2. Then, a series of Sorting Analyzers continue to sequentially sort each tier into smaller subgroups. The observable is chosen to precisely separate the elements of each set in a way that will lead to useful subgroups. The ultimate goal is a specific and unique set of observables that can be applied to identify any single member of the set.

CLASSIFYING ANALYZER

3. The objects that have the observable property are always placed in the box to the right.

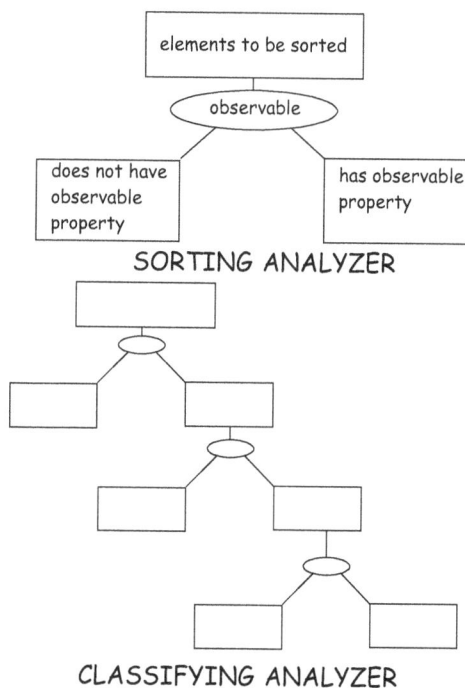

Practice
Exercise

Complete this Classifying Analyzer by sorting and categorizing the elements from the top structure. **The objects that have the observable property are always placed in the box to the right.**

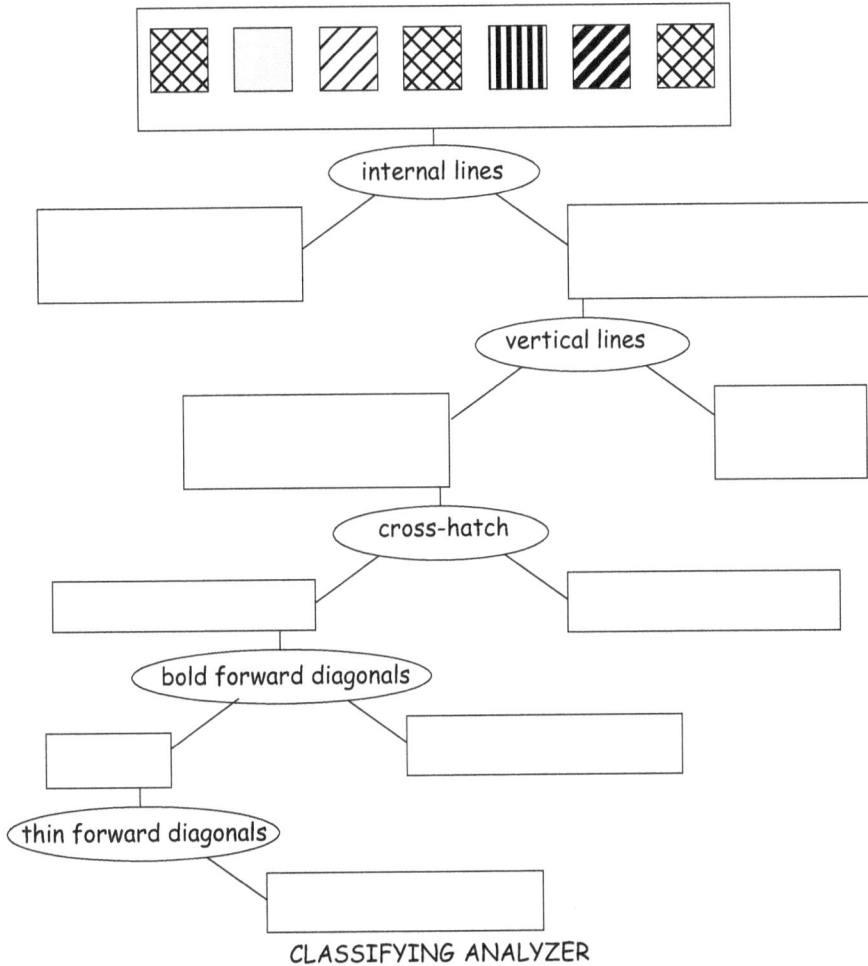

CLASSIFYING ANALYZER

Answer

Here is the answer to the previous exercise. The objects that have the observable property are always placed in the box to the right.

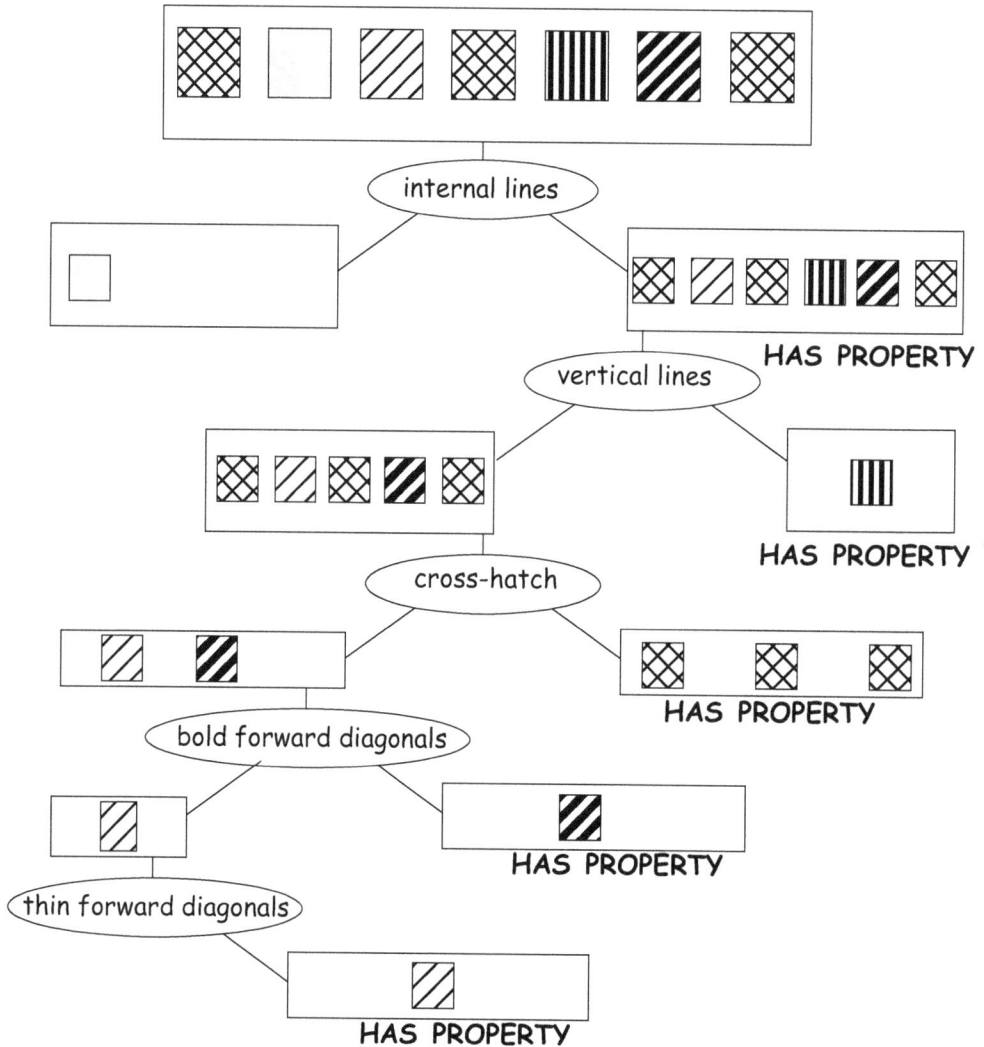

CLASSIFYING ANALYZER

Practice
Exercise

The next example requires some content knowledge of clouds and the weather. Sort the following set of clouds by the observables of precipitation (<u>rain</u>, <u>thunderstorms</u>, <u>steady rain</u>, <u>drizzle</u>). Place the correct observable in the central oval. Place the cloud letter in the sorting bin. The element with the chosen observable is placed in the bin to the right.

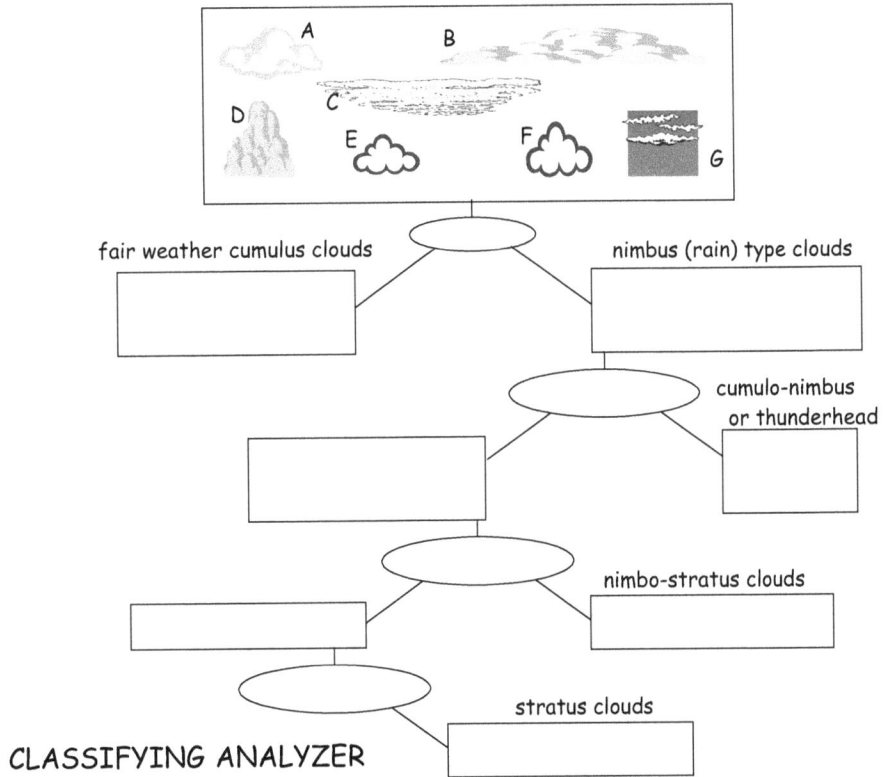

fair weather cumulus clouds

nimbus (rain) type clouds

cumulo-nimbus
or thunderhead

nimbo-stratus clouds

stratus clouds

CLASSIFYING ANALYZER

Answer

This answer to the previous exercise shows how the clouds can be sorted and categorized using precipitation patterns. **Clouds that have the observable property are placed in the right hand box.**

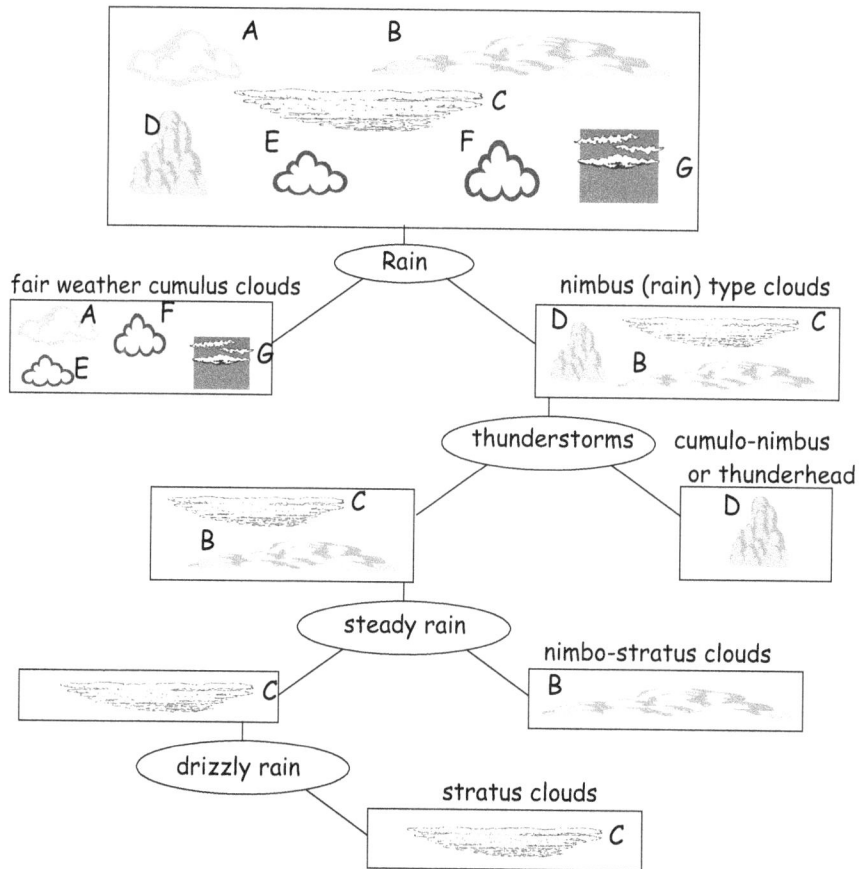

Cross-
Curricular
Example

A Sorting Analyzer can be used outside the science content curriculum. Complete this Category Analyzer using observables inferred in the verse. A variety of answers are possible depending on the initial observable chosen in your analysis.

Jack Sprat could eat no fat
His wife could eat no lean
And so between the two of them
They licked the platter clean.

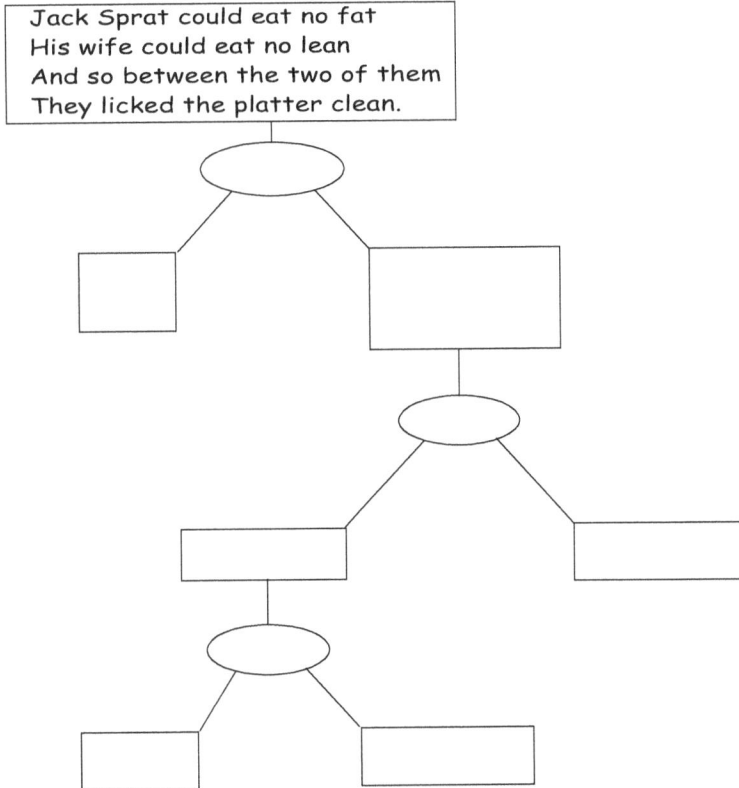

CLASSIFYING ANALYZER

Answer

A possible answer to the Jack Sprat verse is given. Your version may be different depending on the observables that you have chosen.

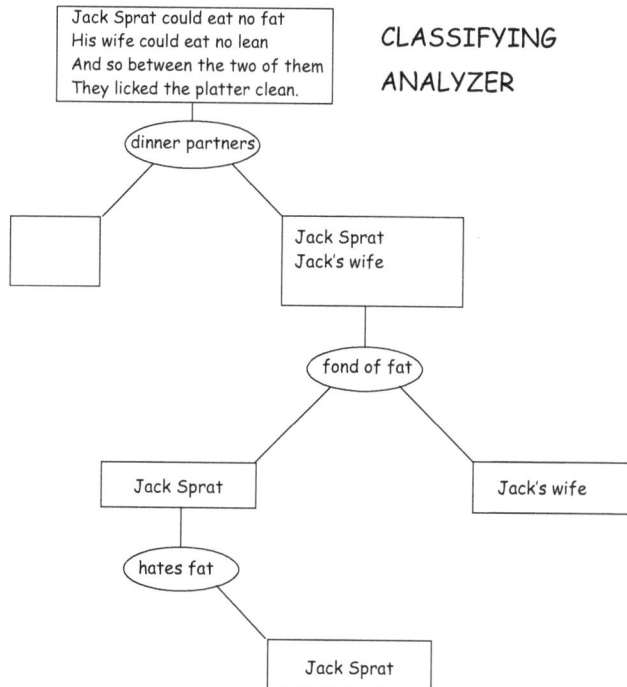

CLASSIFYING

ANALYZER

Summary

In the last two chapters, we have seen how the properties can be used to describe and analyze many aspects of the world. Now we will turn our attention to describing the background space of a system.

The Elements of Systems Modeling

CHAPTER 9

Background Space Graphical Analyzer System

Analyzer	Scientific Process
Background Space Properties Analyzer	• Identify the context in which a system is found
Background Space Feature Extractor	• Define the coordinates and dimensions of a system
	• Extract the properties of background space in order to infer and describe rules in a system

Appreciating the Background Space

The most neglected aspect in exploring the world is an appreciation of the importance of the background space where the elements of a system are found. The background space provides the context or environment of a system. As such, it is a very important aspect of the ecology of a system. Ecology in this context means the overall interaction of each part of a system, not only with other parts of the same system, but with other systems that may also reside in the same background space as well. If the background space is not appreciated, the interaction between systems will not be able to be detected or recorded.

At the end of the twentieth century, people frequently talk about "ecology" without realizing the importance of knowing the details of the background space. Many people would consider that the concept of ecological thinking was raised to the popular mind by books such as Rachel Carson's *The Silent Spring*. *The Silent Spring* described the damage done to birds and other animals by the use of pesticides, notably DDT, that were used to control the destruction of agricultural crops by insects. In the original frame of reference, the background space for the interaction between insects, plants, and humans focused on the role insects played in the destruction of crops and the transmission of disease. A powerful pesticide like DDT offered benefits to the inhabitants of the Earth. However, it was originally unappreciated that pesticides could damage other species in the biosphere, because it was not recognized that the background space also included the ground water cycle and fish! This is a classic case of discovering the background space due to an unexpected cause-and-effect result that moved through a previously unobserved part of the system.

Biosphere contamination by pesticides is not the only example of unexpectedly confronting the properties of the background space. Many scientific advances result from the new perspective that recognizing the previously unobserved properties of the background space can provide.

Making Connections Requires Knowledge of the Background Space

To fully describe a structure, we must be able to extract and describe the rules of arrangement. However, rules can only arrange elements "somewhere". Before extracting the rules, we must know the characteristics of that "somewhere".

The common sense of this is obvious, yet every teacher should be aware that the most common mistake made by both the beginner and the experienced scientist is neglecting to perform a careful description of the background space and to pay proper attention to the properties of that space. One of Einstein's great contributions to modern knowledge (general relativity) was simply the recognition of the properties of the background space that makes up our universe.

Learning About Background Space	Background space is the place where things are put. Learning about the background space has three levels.

1. The first level is easy and accessible to every student even in the earliest grades.

- The Progression of Inquiry and Language of Patterns encourages every critical thinker to start thinking about boundaries and the properties of the space where things happen.
- Most early knowledge of background space will come from considering common experiences like boundaries of playing fields, "inside" and "outside" voices, classroom versus playground behaviors, etc.

2. The next level is the characterization of the background in more abstract terms, including appreciating that the space is one, two, three, or four dimensional (time matters!). Characterizing time in the background space will be discussed in more detail in the next section, Characterizing Background Space.

- This type of characterization is more sophisticated and requires more math and language skills and a developing capacity for inference and abstraction.

3. Finally, a detailed description and mapping of the background space requires sophisticated knowledge and more advanced cognitive skills levels. Students must be able to think metaphorically and use mathematical ideas.

All three levels use the Graphic Analyzer System, but the role that the associated Analyzers will play in a classroom or your work will depend greatly on the students that you teach or the problems that you solve. The crucial recognition of the <u>existence</u> of the background space derives from thinking about it in the first place. This alone is an accessible advance in knowledge for every student at every level!

Background Space Analyzer

The background space includes three parts:

1. A **sample space** where the system itself is found

2. A **boundary** that defines the extent of the sample space

3. The **surrounding space** which is the remaining background space outside of the boundary.

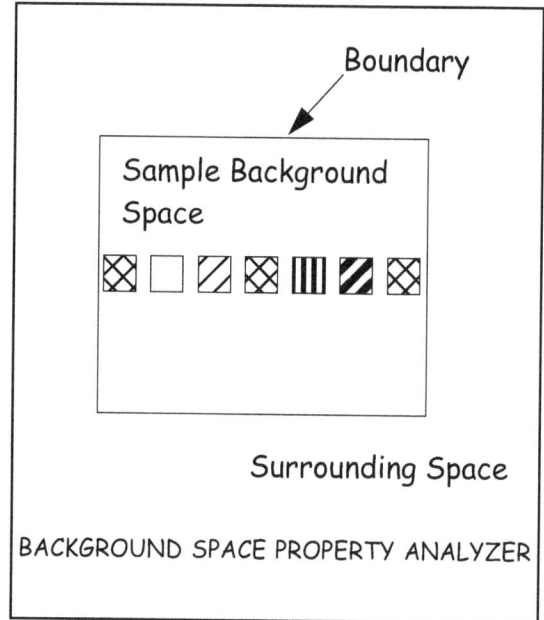

Boundary

Sample Background Space

⊠ ▢ ▨ ⊠ ▥ ▨ ⊠

Surrounding Space

BACKGROUND SPACE PROPERTY ANALYZER

Describing The Background Space

Background space has properties. These properties can be treated just like systems and elements and can be analyzed with a Property Analyzer.

Here are a variety of examples to explore the background space and its influence on understanding a system.

Example 1

What is the description of the background space for a balloon if it is filled with helium?

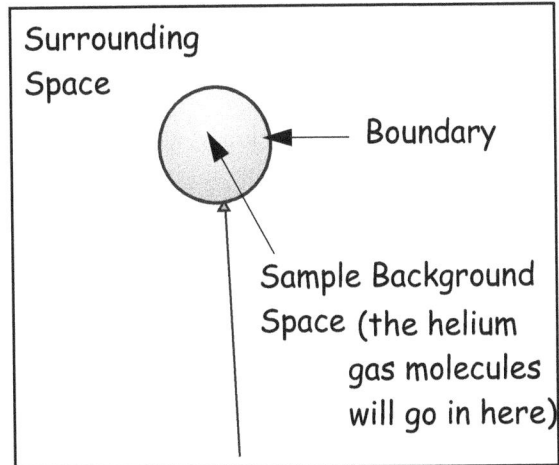

- The skin of the balloon is the **boundary** between the outside and the inside of the balloon.

- The properties of the boundary are that it will not let the helium gas across the boundary.

- The **sample background space** or container space is the space inside the balloon. The **surrounding space** is the space filled by the air outside the balloon.

The properties of the floating balloon depend on all three of the conditions of the background space. As long as the boundary contains the helium gas within the balloon, the inside sample background space will be filled with less matter than the outside atmospheric space; the balloon will be lighter than air and thus will float. The properties of the space inside the balloon and outside are the same, however, the elements that occupy the inside space are relatively less dense than the outside space.

If the balloon bursts, the boundary is disrupted. The helium gas freely escapes from the **sample space** into the **surrounding space**, and the balloon will no longer be less dense compared to the atmosphere. The skin of the balloon will fall from the sky. This explanation of why a helium balloon rises requires an understanding of the background space in which it exists.

Example 2

Learning about backgrounds is easier by considering spaces and boundaries that are commonly appreciated.

The sample background space, boundary, and surrounding space of a playing field are labelled. Are the elements treated differently when they are in these different spaces?

Surrounding Space

Boundary

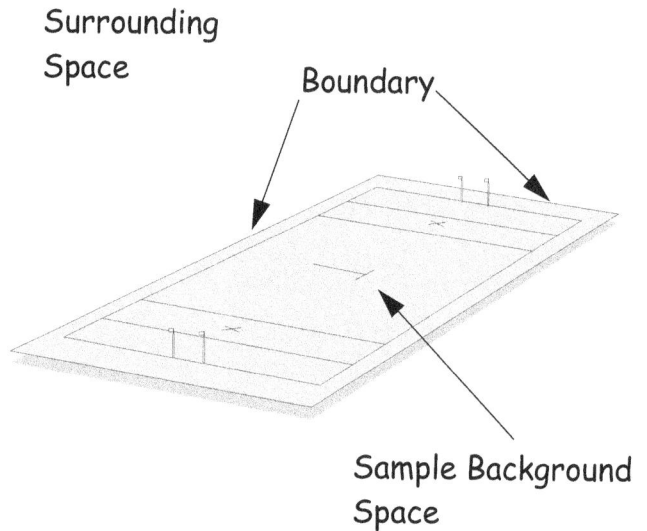

Sample Background Space

The behavior of a player, whether on the field or on the bench, is determined by the boundary of the background space (the sideline).

Practice Exercise

Label the background space (surrounding space, sample background space, and boundary) for this balloon.

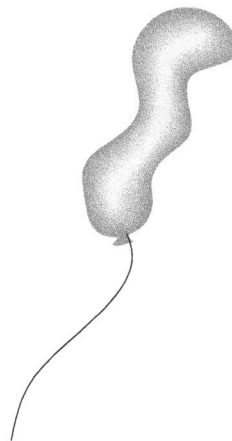

The Elements of Systems Modeling

Example 3

The parts of the background space for a can of soda in a grocery cart are labelled.

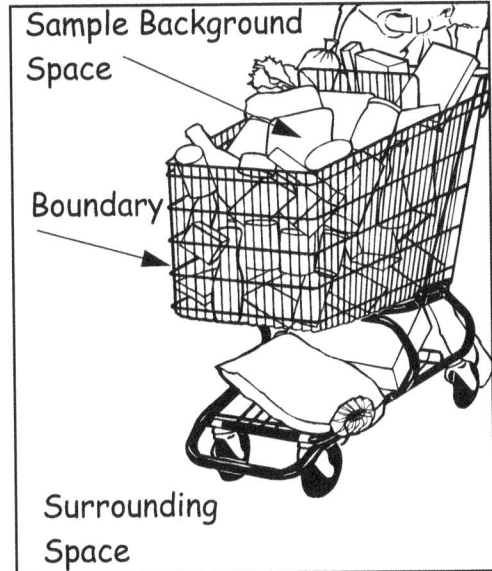

Sample Background Space

Boundary

Surrounding Space

- The overall background space is the space and time in which we all exist. It surrounds the shopping cart.

- The can of soda is arranged in the sample background space of the grocery cart.

- The boundary is the wall of the shopping cart.

- Consider the can of soda inside the sample background space compared to its relationship to the buyer when it sits on the shelf in the surrounding space. Inside the sample space (the shopping cart), the can of soda moves out of the store as a purchase by the shopper, a distinctly different state of existence compared to staying on the shelf in the surrounding space.

- The properties of each of these parts of the background space play a role in the behavior of the elements in the shopping cart and their relationship to one another.

The previous analysis helps us determine the qualities of the background space and helps us define what the boundaries of interest are in a system. A qualitative description of the space of the system, the surroundings, and the boundaries are all very important in understanding a system's behavior.

Practice
Exercise

Label the background space (surrounding space, sample background space, and boundary) for this squash court.

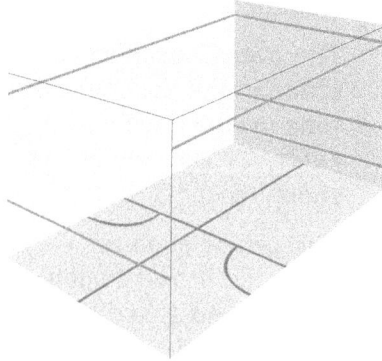

Example 4

The perpetual motion machine is an example of a background space misconception. Perpetual motion machines appear to violate the law of nature that says that energy can not be created. Measurements of energy production and use inside the sample space make it seem that more energy is being produced than used. However, all perpetual motion machines draw energy across the boundary from the surrounding space. Therefore, it appears that they are making energy when in fact they are getting it from an unobserved part of the background space.

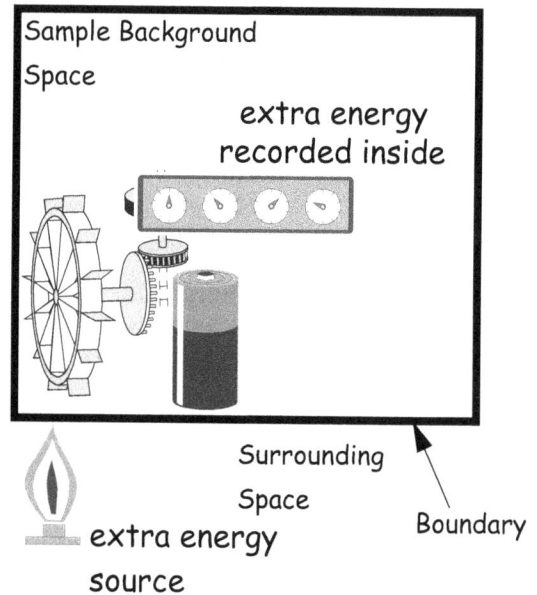

Sample Background Space

extra energy recorded inside

Surrounding Space

Boundary

extra energy source

Answers to
Previous
Exercises

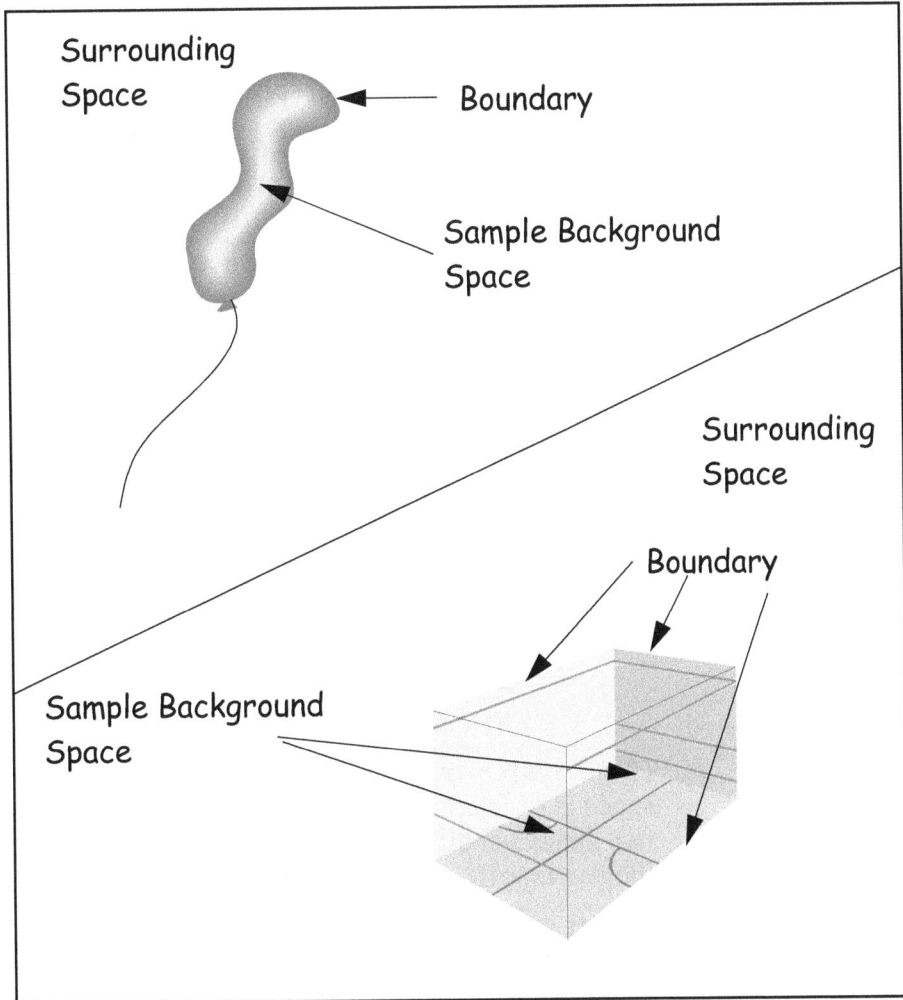

Surrounding
Space

Boundary

Sample Background
Space

Surrounding
Space

Boundary

Sample Background
Space

Sample Background Space	In order to determine and describe the character of the background space with the precision necessary to form a complete map, we must examine the more restricted portion of the background space (the **sample background space**) that contains the structure of interest.

- The easiest way to do this is to define a "container" space.
  The container holds the system of interest.

· The container's properties are then extended to the rest of the space. In this fashion, we can attribute the properties of the sample background space to the greater background space.

A description of the shape of the space and a map of the space is what a Background Space Analyzer provides.

Character-izing Background Space

To characterize background space, we must:

1. Figure out the shape of the space that contains the elements. For example:

· A page in a book (the space in which these words appear) is flat with length and width.

· The Earth's surface (the space on which towns and landmarks are found) is a curved spherical surface

2. Have a sense of the size of the sample and surrounding space.

· A system or structure is located in a background space that is large enough to contain it.

If the system is a solid object like an apple, then the background space is at least three-dimensional (3-D). Words or drawings are two dimensional, since they are contained in the plane of the page (having **length** and **width** only).

Almost every system of real interest has a time dimension (a past, present, and future). For a system to undergo change or evolution, it must move through time. Therefore, most background spaces are actually four dimensional (**length, width, height,** and **time**). When the time dimension is explicitly considered, a **dynamic** system or **dynamic** modeling results.

Mapping Background Space

Measuring background space becomes important when we want to figure out the rules that arrange elements. In order to place elements into the background space, we need to be able to picture it and draw a map of it. If the background space can be drawn, a coordinate system can be defined.

In addition to the shape and size information described earlier, a map of the space requires that:

- A system of gridlines, much like a street map, is applied to the space.
- Once the streets are laid down, each space where an element can go is labeled.

A map of the background space is generally required to make precise models for testing with the scientific method.

The Background Feature Extractor

Container, Origin, Places

The procedure to map the background space for a language-of-patterns analysis is given by the letters COP.

- Container
- Origin
- Places

We will use the filled square pattern structure as an example of the process.

Container

1. Imagine a container (sample boundary space) [A] to hold the system. The properties of the container should be the properties of the background space. Our structure is a set of square boxes with height and length arranged in a line. A two dimensional rectangle will contain the sample boundary space accurately.

- Draw a rectangle around the system.

- Since a rectangle has length and width, draw a line for each of these [**B**]. (In technical terms, this step defines the coordinate system of the space.)

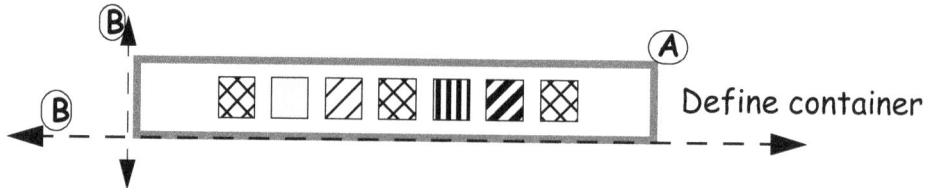

Define container

Origin

2. Next we assign an origin point [O] in the container [C]. This is where we will start drawing the map. Notice that the container forms a boundary around the sample boundary space and the surrounding space.

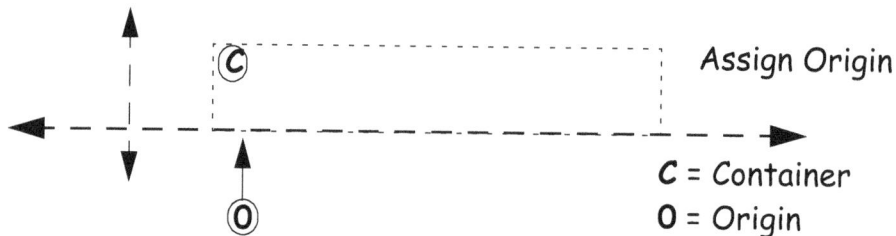

Assign Origin

C = Container
O = Origin

Places

3. Finally, we mark the potential background spaces by drawing background placemarker [**D**] spaces in the container.

- We draw them to a convenient size, in this case the size that will hold the squares. In this example the placemarker spaces are represented by:

4. To find our way around we must label the placemarkers [**E**] just like naming streets on a map.

5. We remind ourselves that the container space is just a sample portion of the overall background space by extending the places marked with dotted lines [**F**].

C = container
D = placemarker spaces
E = placemarker
    number

F = background
    placemarkers
    outside container

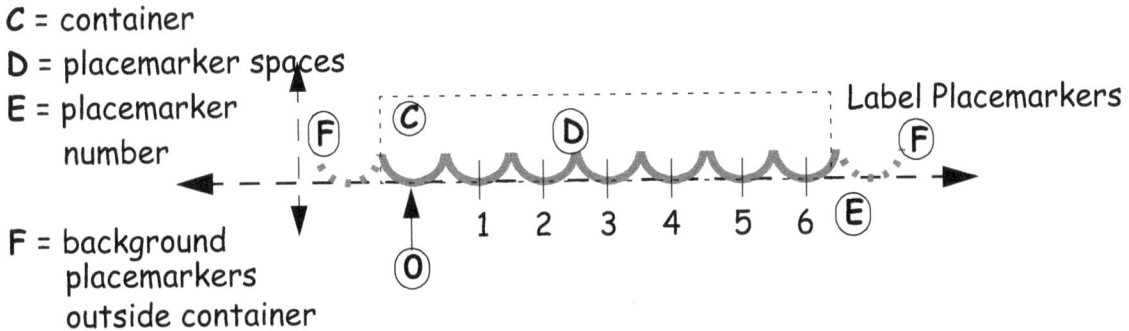

Label Placemarkers

6. With the **container** (and its coordinates), the **origin**, and the **places** established, the sample space is redrawn leaving out the imaginary container. The background space is now described and can be explored itself or can be used for the mapping of rules and the evolution of the system.

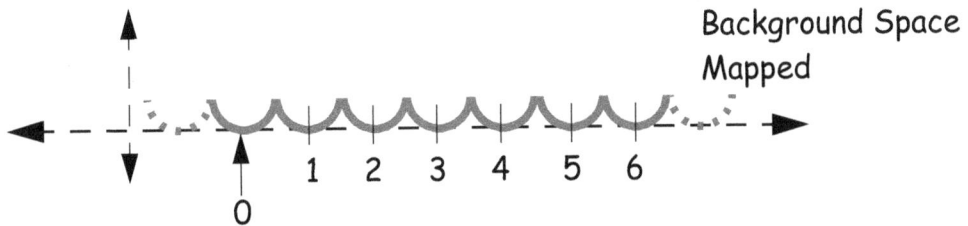

Background Space Mapped

In Summary

An overall graphical summary of the background space mapping process is shown here:

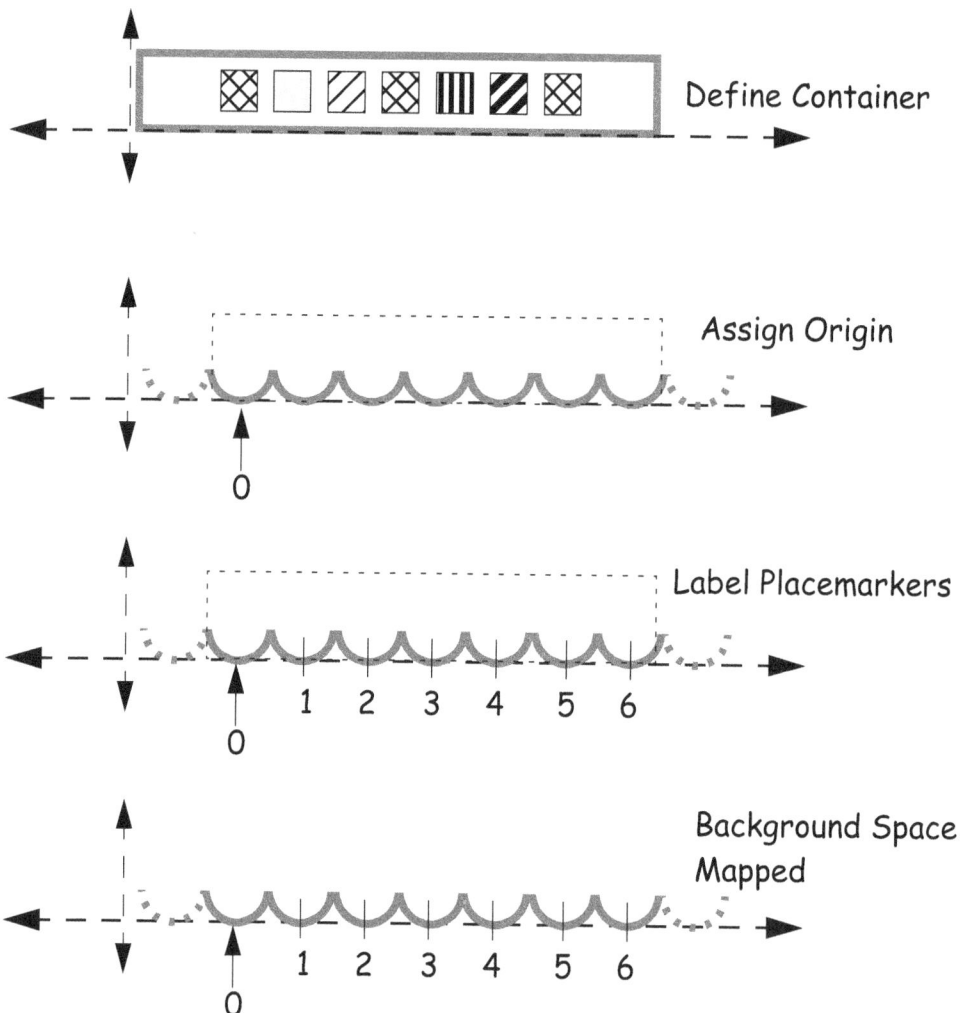

Define Container

Assign Origin

0

Label Placemarkers

0   1 2 3 4 5 6

Background Space Mapped

0   1 2 3 4 5 6

Let's practice extracting the background space properties of some other systems.

- CAUTION: In this process, we frequently find assumptions that we have made to be incorrect. Always keep an open mind to the actual properties that you see.

The Elements of Systems Modeling

Practice
Exercise

Extract the background space of this system. Remember that placemarkers can be a set of coordinates. (Answer on next page.)

System

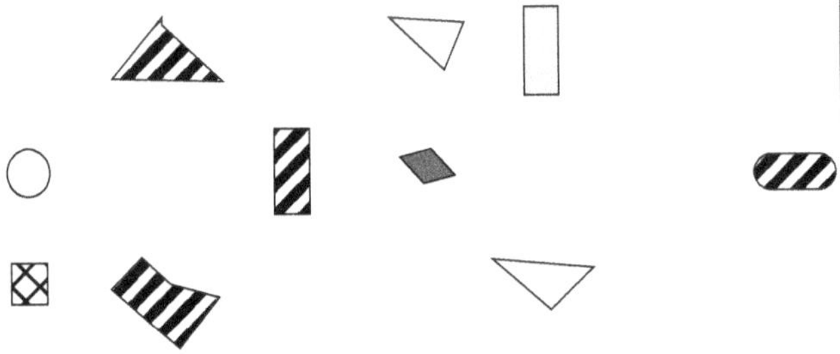

Container

Origin

Places

Answer:
Container

The first assumption is that this drawing exists in the flat plane of the page. Proceeding from that assumption, we draw a container and place the reference lines (axes) that make the container.

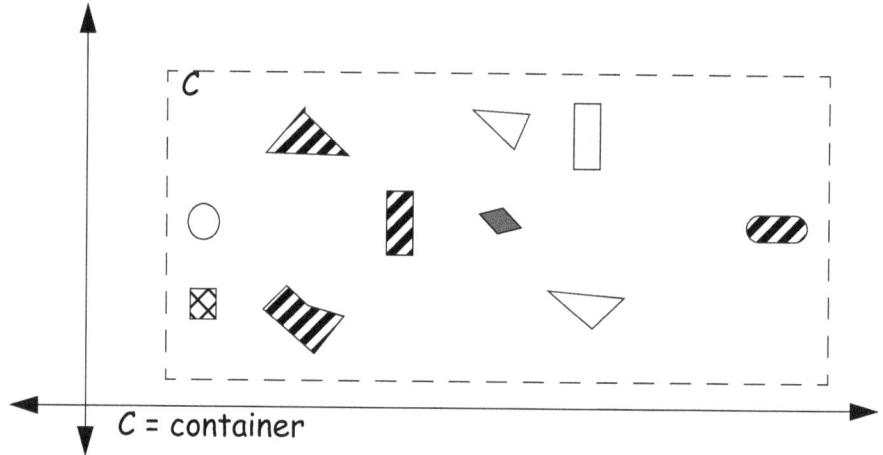

C = container

Answer: Origin   Next we choose a convenient origin.

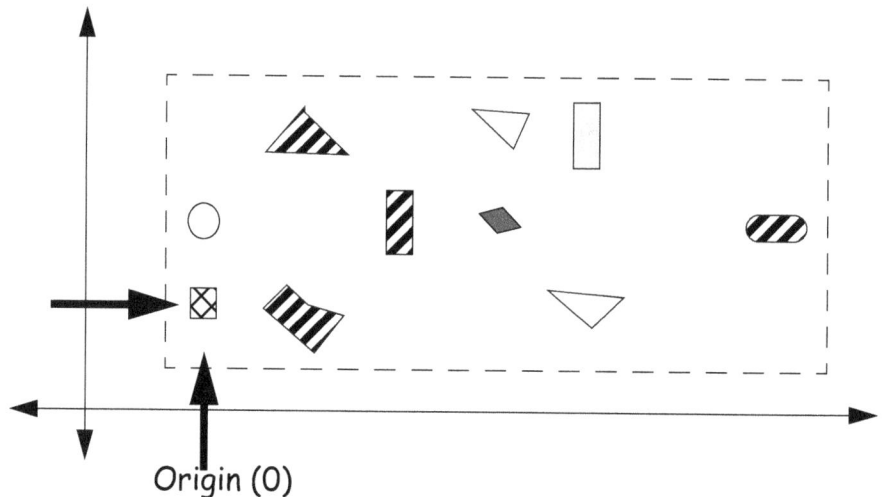

Origin (0)

Answer: Places   Finally, we draw in places to put things, placemarker spaces. Remember to extend the map with dotted lines into the background space beyond the sample boundary. [In solving this example, a Cartesian coordinate system in two dimensions has been drawn!]

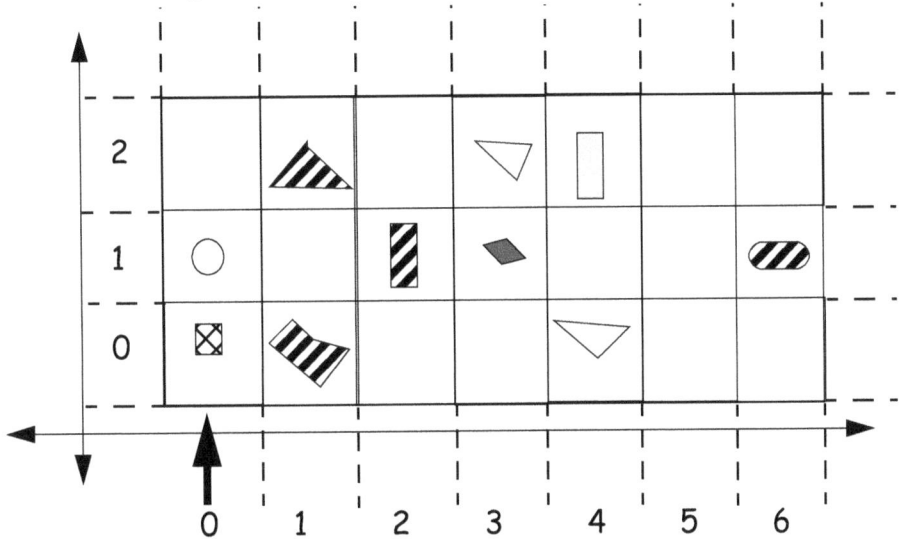

Answer:
Summary

This description of background space is two dimensional in the plane of the page, due to the arrangement and properties of its elements (flat drawings). The location of each object can be identified by its coordinates.

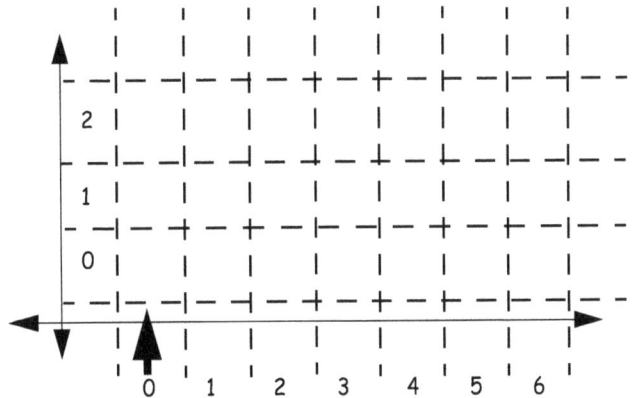

Notes

CHAPTER 10

Exploring Rules with the Graphical Analyzer System

Analyzers	Application
Ordering Analyzer	• Guide inferential process of extracting rules of arrangement by examining the ordering of elements
Rule Extractor	• Extract the properties of background space in order to infer and describe rules in a system • Define the coordinates and dimensions of a system

Arranging the Elements

The extraction of rules is an inferential process. Elements are arranged in one of three ways in a background space. They can be sorted based on an observable into each of these arrangements:

- Nominal arrangement
- Ordinal arrangement
- Spatial arrangement

Nominal Sorting

Elements are always either <u>in</u> or <u>out</u> of the background space based on an observable. This is called a **nominal** arrangement. Nominal arrangements are the same as grouping and sorting and use the Sorting Analyzer.

Ordinal Sorting

Each element can be placed in a sequence based on an observable such as size, age, weight, and time. This is an **ordinal** arrangement. Ordinal arrangements are common and include food chains, sequences, and growth patterns. They require an **observable** and a **measurable** value associated with the observable that can be sequenced.

Practice Exercise

Tadpoles have tails and frogs do not. Tadpoles lose their tails in proportion to their bodies as they turn into frogs. Propose an order for these tadpoles based on the observable of tail/body length. Answer on the next page.

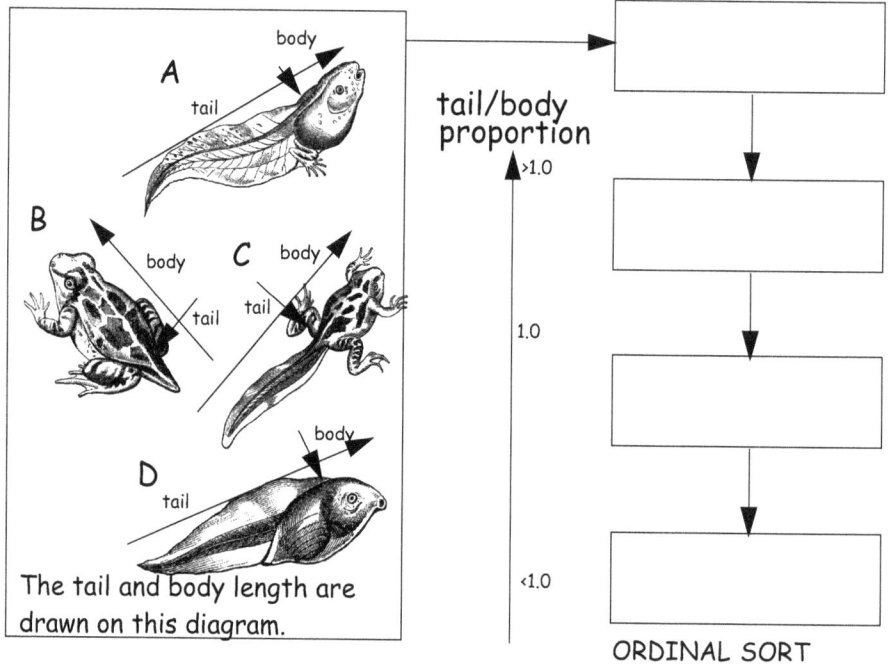

ORDERING ANALYZER

Answer

Answer to tadpole ordinal sort.

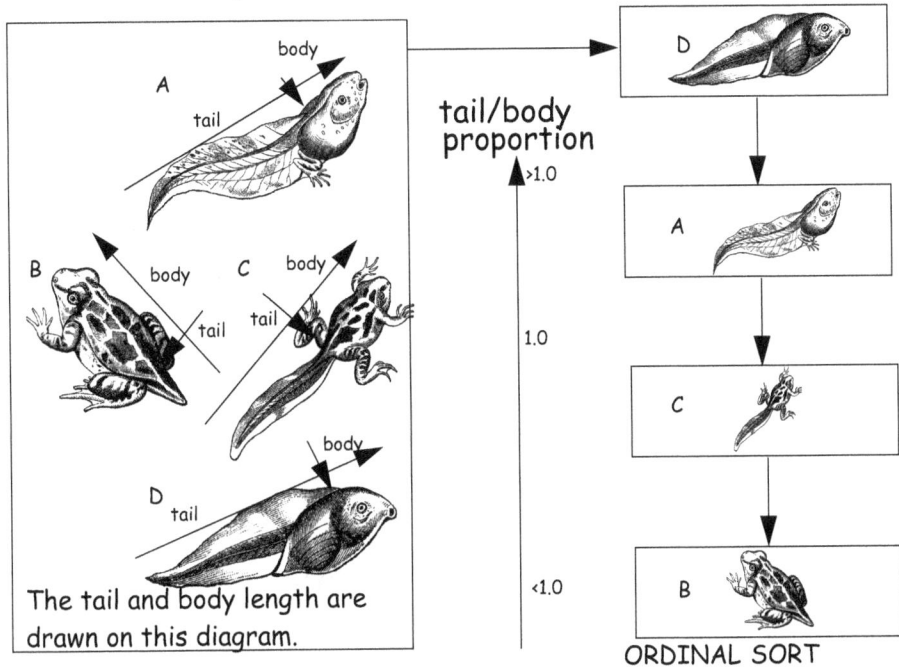

The tail and body length are drawn on this diagram.

ORDINAL SORT

Spatial Ordering

Each element can be placed at a specific point in space based on the background map coordinates. This is a **spatial** arrangement. Ordering elements spatially is a physical mapping that places each point of an object in a space defined by a grid. There must be a measurement for each coordinate of the background space. Spatial ordering uses standard graphing techniques.

The Rule Extractor

Inferring rules is a relatively advanced capacity that becomes available to students in middle school and develops in sophistication through the highest grades.

· A rule operates on an element.

- A rule is always defined in terms of the element on which it is operating.

- Rules arrange elements in some type of space.

- Rules either order elements with respect to some point in the background space or else they order elements relative to another element.

    - Ordering elements with respect to a constant origin in the background space is easier to learn.
    - We will only consider this easier case.

Rules are almost always inferred from the arrangement of the elements in a carefully defined background space. This is why the elements and the background are extracted first. Rules can be extracted by mapping a variable onto a coordinate system.

Start, Move, Repeat (SMR)	The procedure to define the rules for a language-of-patterns analysis is given by the letters SMR.

- Start
- Move
- Repeat

To extract the rules we must always come prepared with:

- The structure
- The elements in the structure
- The extracted background space

We will use our familiar filled-square pattern as an example of the process.

The elements have already been extracted with the Element Analyzer.

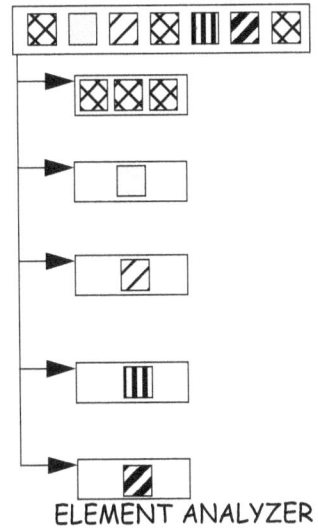

ELEMENT ANALYZER

We have also already extracted and described the background space:

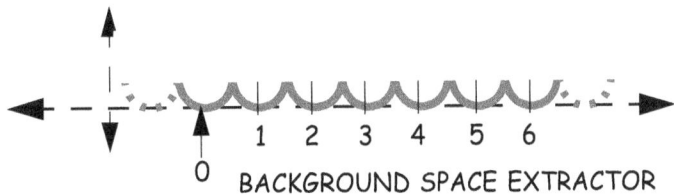

0   BACKGROUND SPACE EXTRACTOR

Now, we can extract the rules for the first element:

1. Define where in the background the element **starts (S)**.

2. Determine where the element **moves (M)** (what direction and how far).

3. Determine how often the move is **repeated (R)**.

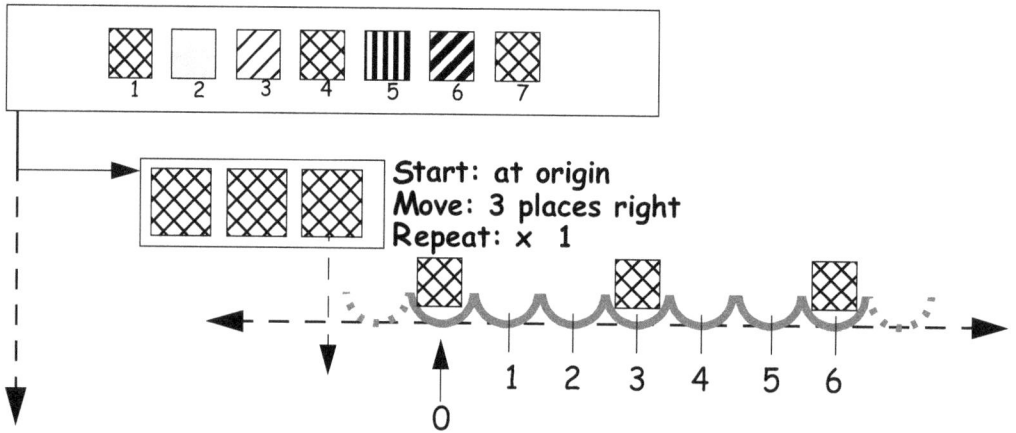

Start: at origin
Move: 3 places right
Repeat: × 1

Now, we extract the rules for the second element:

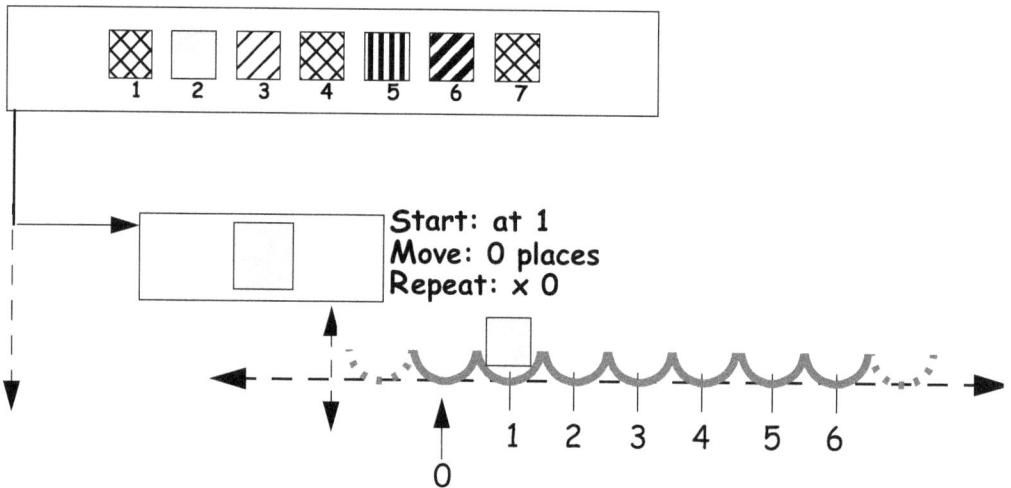

Start: at 1
Move: 0 places
Repeat: × 0

The rules arranging each element can be sequentially extracted in this fashion.

Practice
Exercise

Complete the extraction process for elements 3, 4, and 5. Write the rules in the SMR space. A summary of the rule extraction for this pattern of elements is shown on the next page.

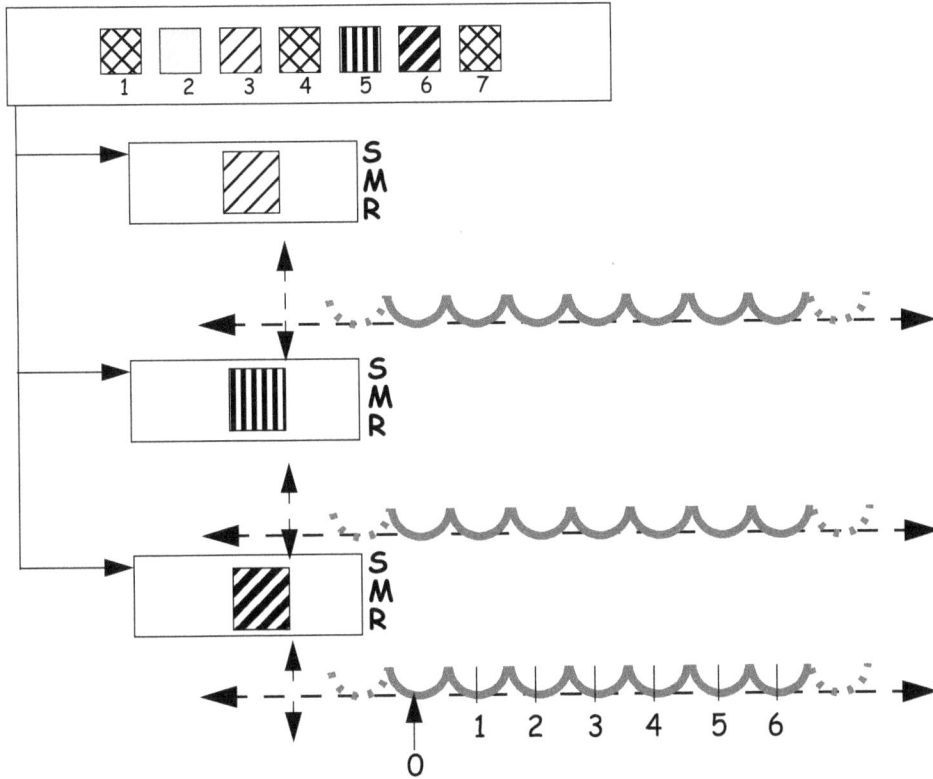

Answer

Completed rule extraction for the fill-pattern structure.

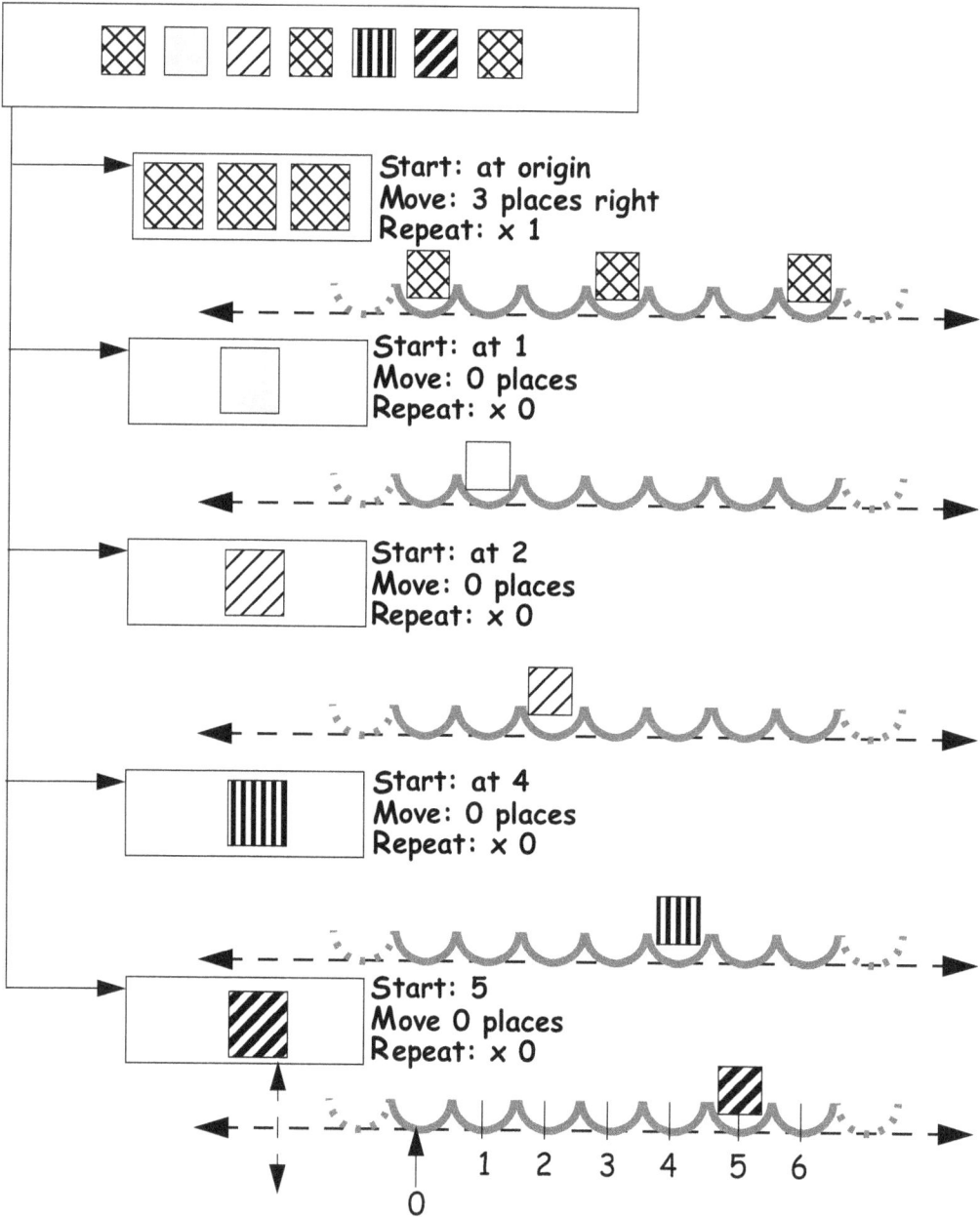

Start: at origin
Move: 3 places right
Repeat: x 1

Start: at 1
Move: 0 places
Repeat: x 0

Start: at 2
Move: 0 places
Repeat: x 0

Start: at 4
Move: 0 places
Repeat: x 0

Start: 5
Move 0 places
Repeat: x 0

0   1   2   3   4   5   6

This Rule Analyzer is very simple to use in terms of step-by-step processes, but it can get complicated as the structures grow in size, element categories, and dimension. Remarkably, this process is sufficient to extract the rules and write them clearly and precisely for the majority of the systems likely to be encountered in the natural world.

For example:

  - In biochemistry and medicine, the determination of the shapes and functions of proteins and enzymes are determined by characterizing the arrangements of atoms extracted from the pattern of x-ray crystallography.
  - The complex and complicated spectral fingerprints of stars and chemical compounds are described and interpreted by understanding the rules that arrange their observables of light, color, and chemical reaction.

Patterns and Rule Extraction

Generally, when we talk about patterns, we are referring to a certain type of structure in which certain elements are following a repeating rule of arrangement.

  - Finding the pattern in a structure is a rule extraction.
  - This is why describing a pattern in terms such as ABABAB defines the kind of repetition that an element will have, but it does not describe the overall composition of the structure.
  - The Language of Patterns can describe the repeating rules, elements, and backgrounds of a typical pattern, as well as efficiently describe any system of elements or structures.

Practice
Exercise

This example **is** rocket science. Extract the rules of this rocket's motion. Here is a hint: the motion of a rocket is an arrangement of positions of that rocket.

Answer

The rocket's motion is described by the sequential arrangement of its position as it moves through space.

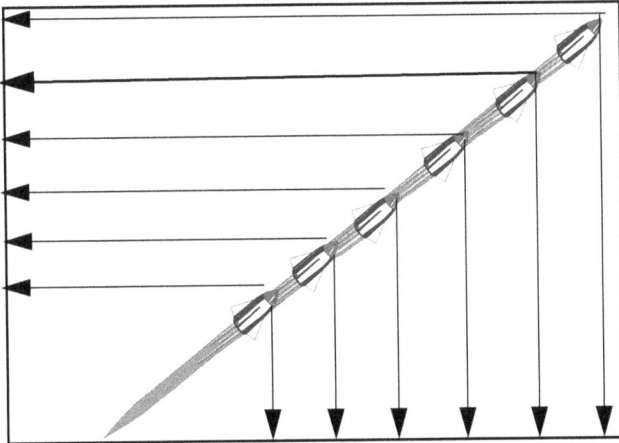

The elements of the motion are six points in a horizontal direction and six points in a vertical direction

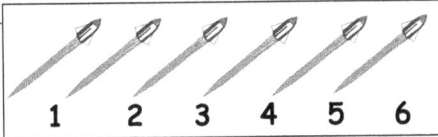

1   2   3   4   5   6

The background space is a square container with a vertical (y) and horizontal (x) axis.

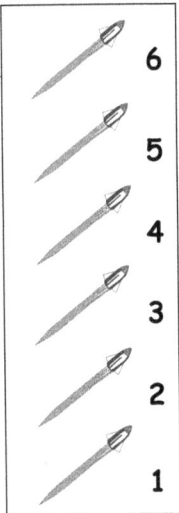

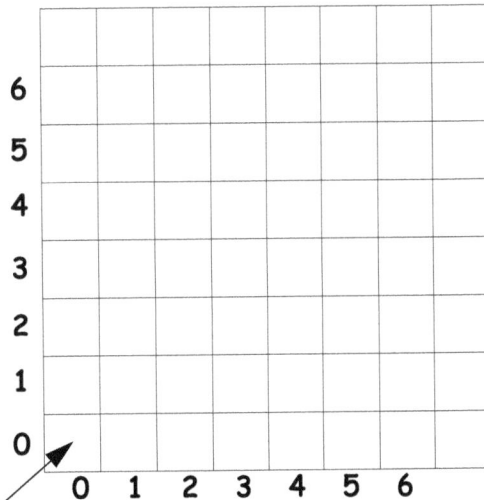

6
5
4
3
2
1

6
5
4
3
2
1

0
0   1   2   3   4   5   6

Origin

The origin is chosen and the placeholders are a grid on the x and y axes

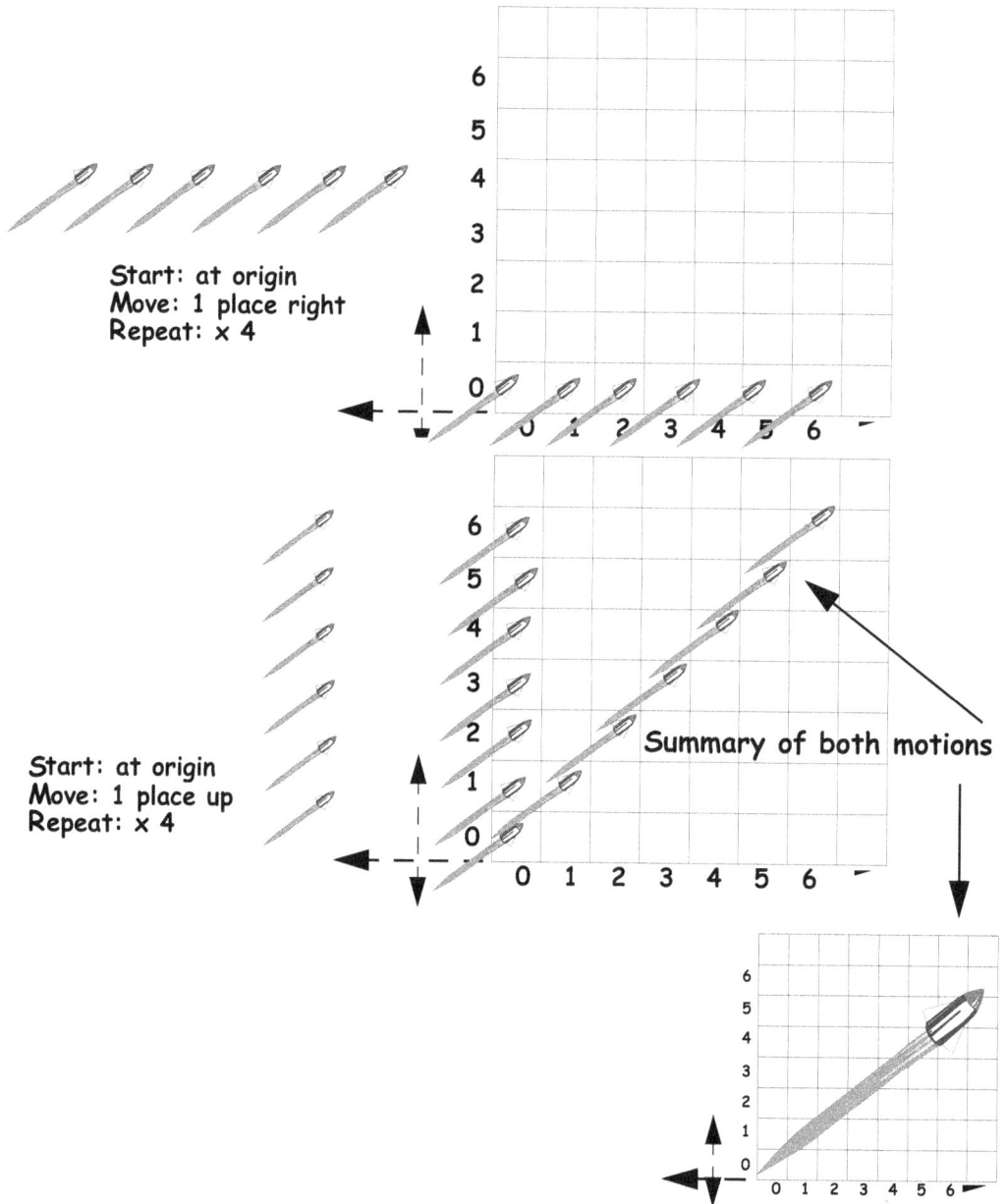

Start: at origin
Move: 1 place right
Repeat: x 4

Start: at origin
Move: 1 place up
Repeat: x 4

Summary of both motions

CHAPTER 11

Using the System and Structure Graphical Analyzers

Extending Systems Description

Analyzers	Application
System Analyzer Structure Analyzer	• Characterize the composition of a system or structure in terms of its elements, rules and background space

Describing a Structure

Describing a system or a structure requires knowledge of the elements, rules, and background, as well as how they go together.

The separation of a system into its parts is a concrete activity that can be performed by students as early as first grade. As the student's skills grow and the learner matures cognitively, the same analyzer tools can be used for more sophisticated exploration. This includes:

- The contrast and correspondence between the organization of systems.
- The inferred knowledge of how a system is organized.
- The ability to use one system as a metaphor or model for another.

Defining Systems

When elements are arranged in a particular background space by a set of rules that describe the relationships among the elements, background itself, and other rules, we have a **system**. Systems give rise to emergent properties and can be arranged in a larger space by rules that apply to a level of systems such that they can be arranged to form larger systems. So at one level, systems can be treated like elements in a larger system, i.e. systems are formed from sub-systems arranged in space.

The power of the Language of Patterns is the use of the same critical process to find connections in systems of objects and ideas. The same process that allowed us to describe how things go together can be used again and again until we see all the parts and relationships both at a smaller and at a larger level.

Example

Consider the systemic structure of a very familiar game such as kickball.

· Make a list of what is needed to play kickball.

> · First, we need to get the players together. Ideally, we need 12 -18 people.
> · We need a kickball.
> · We need four bases.
> · We need a place to play.
> · We are going to have to pick two teams from our players.
> · When a team takes the field, it will have a pitcher, a first, second, and third base player. Each team needs a catcher and at least one or two outfielders.
> · Another element of the game will be the number of runs gained by each team as they play. This will be kept as a tally called the score.

Kickball list	
Billy	Suzie
Linda	Sam
Joshua	Peter
Cindy	Catherine
Chris	Tom
Barbara	Earl
four bases	
kickball	
field on Vine street	
catcher	
second base player	
first base player	
third base player	
pitcher	
score	

- The structure of a kickball game can be drawn graphically on the Structure Analyzer:

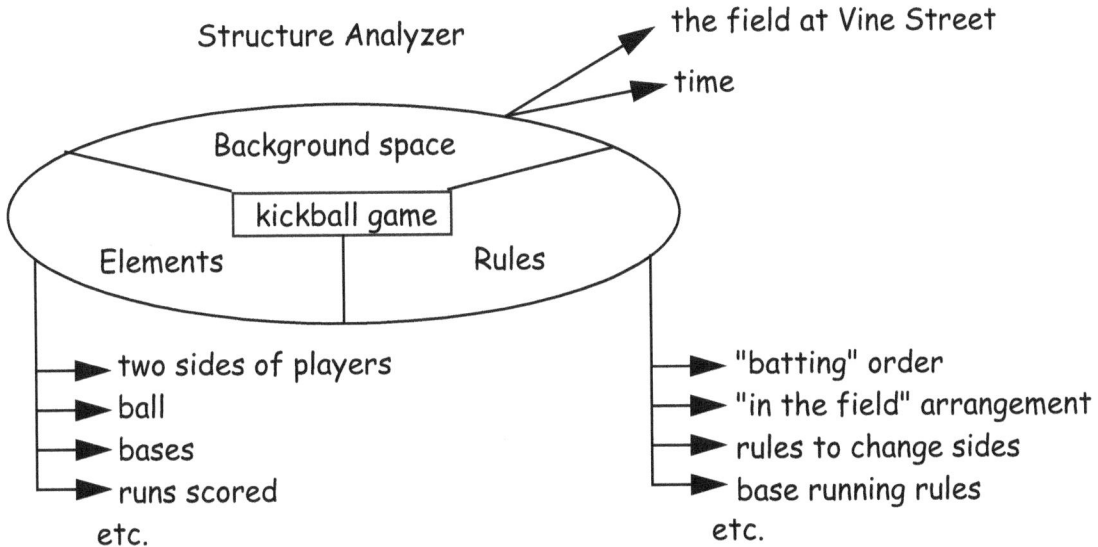

Structure Analyzer

the field at Vine Street

time

Background space

kickball game

Elements

Rules

- two sides of players
- ball
- bases
- runs scored

etc.

- "batting" order
- "in the field" arrangement
- rules to change sides
- base running rules

etc.

- This kickball game has a structure that includes:

    - A background space - a place and time to play.
    - Elements - two teams or sides, a ball, bases, and a scoring system.
    - Rules - arranging all of the different elements into batting sides, fielding sides, ways to change sides, rules for keeping score and then determining who wins, etc.

Each of the components of the structure (elements, rules, and background) have their own structure which can be described by elements, rules, and background.

Let's consider just the ball.

    - The ball is an essential element of the game.
    - It has certain properties that define it as a kickball.
    - These properties are the result of its structure.

Right now, lets just be concerned with the properties that describe the kickball. Either the Property Analyzer or the Element Analyzer can be used for this purpose.

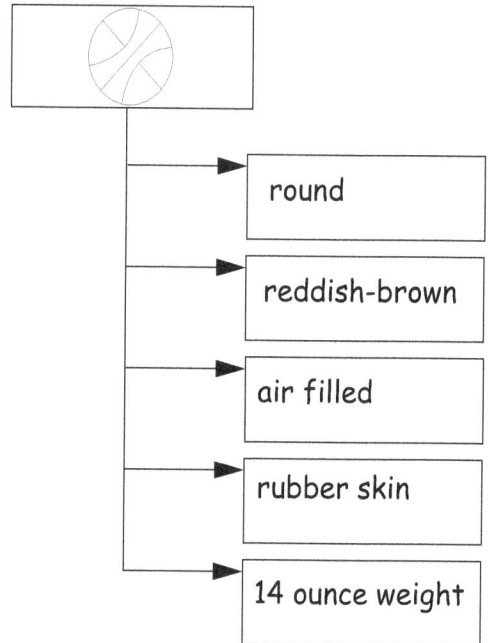

ELEMENT ANALYZER

Element analyzers can chart properties to whatever degree of precision needed for the topic of interest.

Most of us may not need to know or care about the actual measurement, materials, or air pressure need to make the kickball regulation, but they are properties of the ball none-the-less.

The background space is very important for the structure of the game. Just a couple of examples can serve to highlight this. First, we must consider the properties of the background space.

- Is the kickball game to be played after school on a night with a lot of homework?

    - Then the game will be limited in terms of its length, because the player elements will have to go home to do their homework (if they are allowed to play at all!).

    - A Saturday game might have no limit on time, but the sun might be in the eyes of the kicking team in the morning hours, leading to very different games on the weekend versus after school.

    - The same rules and players, and elements are all affected by the background space or context of the game, in this case the time of the game.

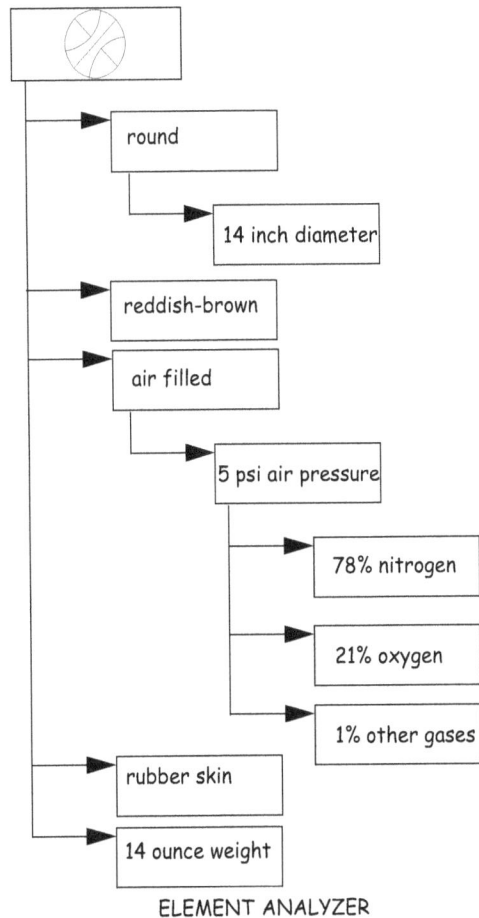

round

14 inch diameter

reddish-brown

air filled

5 psi air pressure

78% nitrogen

21% oxygen

1% other gases

rubber skin

14 ounce weight

ELEMENT ANALYZER

- What if the game is played on a field with woods behind the outfield versus a city sand-lot with a brick wall at the back of the outfield?

    - A kick out of the outfield and into the woods will surely be a home run.

    - However, a kick out of the outfield into the brick wall can be caught on the fly and may be an out!

    - The same rules and the same players again are affected by the background space.

- Finally, the background space determines what is in-bounds and what is out-of-bounds. If the ball falls inside the boundary line, one set of rules apply to the players and ball. If the ball falls outside the boundary line, play proceeds differently.

Background space and how it is structured have a huge impact on the way the game itself will be structured.

CHAPTER 12

Evolution and Change with Graphical Analyzers

Analyzers	Application
Evolution and Change Analyzer	• Identify and describe the part of a system that causes it to rearrange and exhibit changed properties

More on Systems and Structures

We have been talking about **systems**, **structures** and **elements** throughout this Tour.

- Usually, we think of an element as being a simple object, assembled into a larger and more complicated structure. We might think of a nail as being an element of a table. The relatively complicated arrangement of wood, screws, nails, and brackets is usually thought of as a structure. However, if the table is viewed as part of a much larger system, a cafeteria for example, it becomes an element of the system of tables, chairs, food, waiters, and customers.

- Alternatively, the chemical and physical organization of the nail itself is a highly complicated arrangement of smaller chemical elements. The arrangement of these smaller objects constitutes the structure of the nail.

- The following example helps us to appreciate that each element of a system has its own internal structure and its own properties and is therefore a (sub)system.

- In general, elements, rules, and background spaces each have properties and can be analyzed as if they were (sub)systems themselves. Analysis of any of these can be done by the Language of Patterns. The essential choice of the observer is to decide how closely the system or object needs to be examined.

- Systems can change when some force changes their elements, rules, or background space. A new system is the result. A method of mapping these changes is our concern now.

Evolution or Change Analyzer

Structures and systems undergo change. When they change, they typically develop new properties. These new properties are the result of a specific change to some aspect of the original system. The process of change must be described in terms of the alteration of the original elements, arrangements, or background space. Often systems change over time. As was mentioned earlier change with respect to time are called dynamic change.

For example, if we <u>changed the space</u> upon which our now familiar series of squares lays from a within a rectangle to a donut, <u>the system and its structure will change</u>.

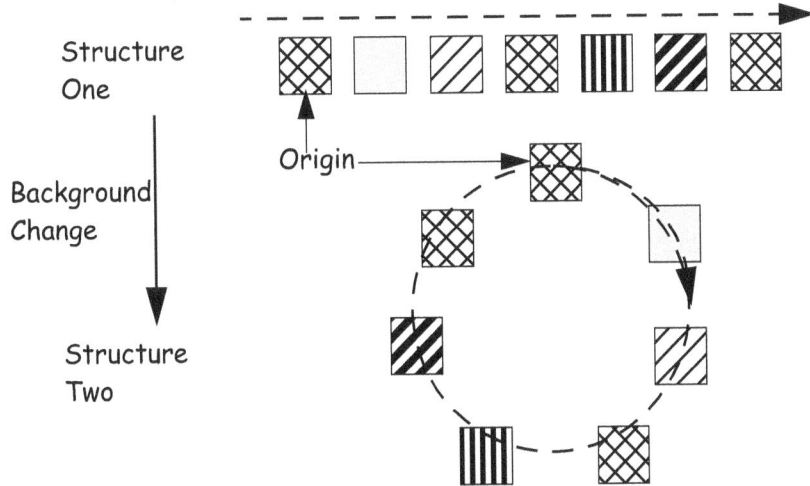

Structure
One

Background
Change

Structure
Two

Origin

The elements are the same (seven textured squares). The rules are the same (identical order and arranged along a central line through the center of each square). Most of the properties of the background space are the same (everything is still in the plane of the page). Yet by changing the shape of the background space (a circular rather than a linear space), the structure and the properties of the system are very different.

The Structure Analyzer can be used to describe the linear structure:

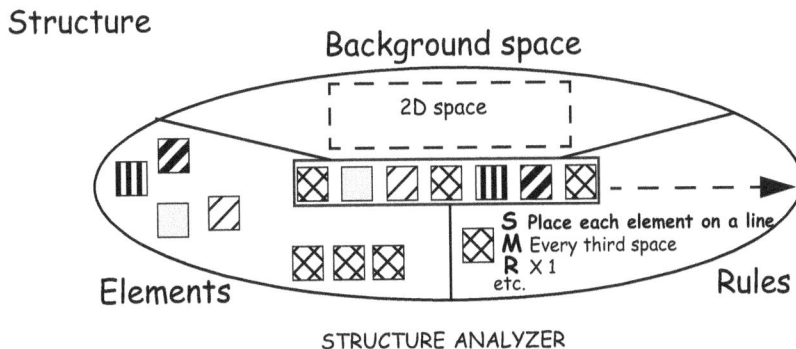

Structure

Background space

2D space

Elements

S Place each element on a line
M Every third space
R X 1
etc.

Rules

STRUCTURE ANALYZER

We can describe the change in structure from a linear structure into a circular one by using a Change and Evolution Analyzer.

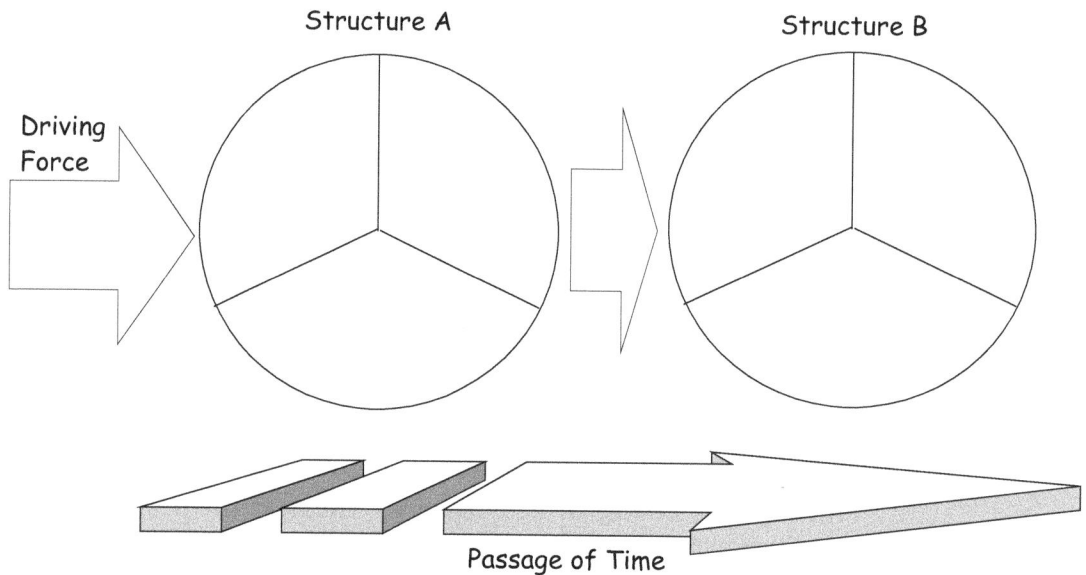

Structure A   Structure B

Driving
Force

Passage of Time

CHANGE AND EVOLUTION ANALYZER

Change (or evolution) occurs when some force acts on one of the elements, rules, or background of one structure to make it change into another structure.

In the case of the evolution of our "fill-pattern", a force acted to change the surface for "unfolding" the elements. Instead of having the order of the elements unfold on a linear surface, the surface changed, and the elements unfolded onto a curved surface:

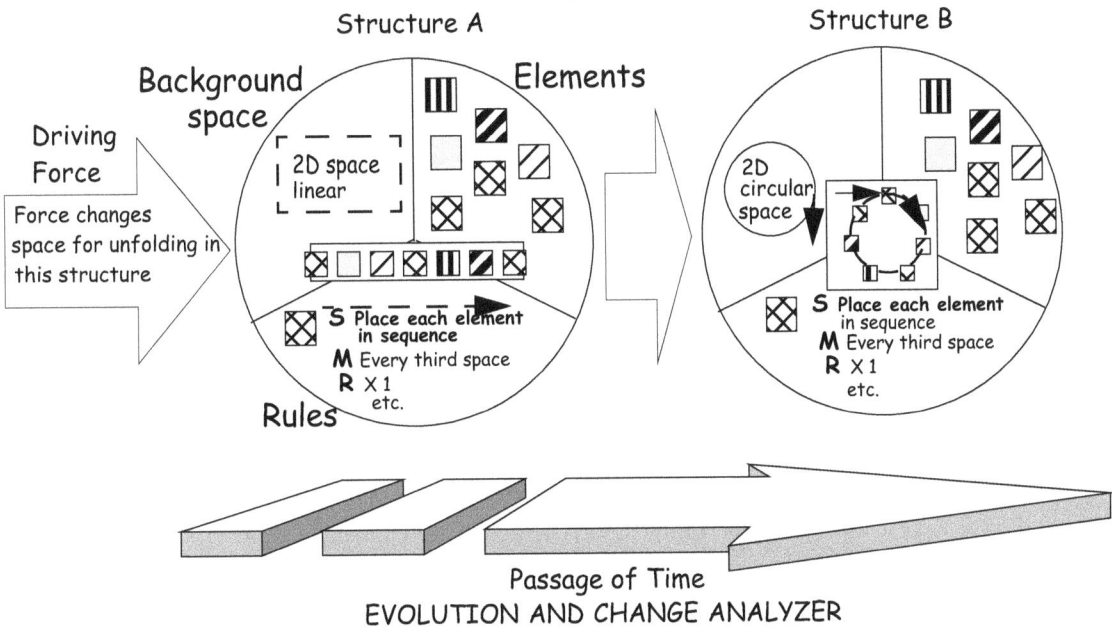

EVOLUTION AND CHANGE ANALYZER

Both structures are similarly ordered and share many of the same properties, but clearly they are different structures with different functions.

Example

What happens to a stable pond ecosystem when a drought occurs?

First, the structure of the pond must be described. This is done using a Structure Analyzer.

The force acting on this pond is a severe drought in which the region is over 20 inches behind in rainfall (the conditions in the Northeastern USA in 1999).

Where does this force have its effect? The background space is not changed nor are the rules. However, the drought reduces the water table and the amount of water entering the pond by rainfall.

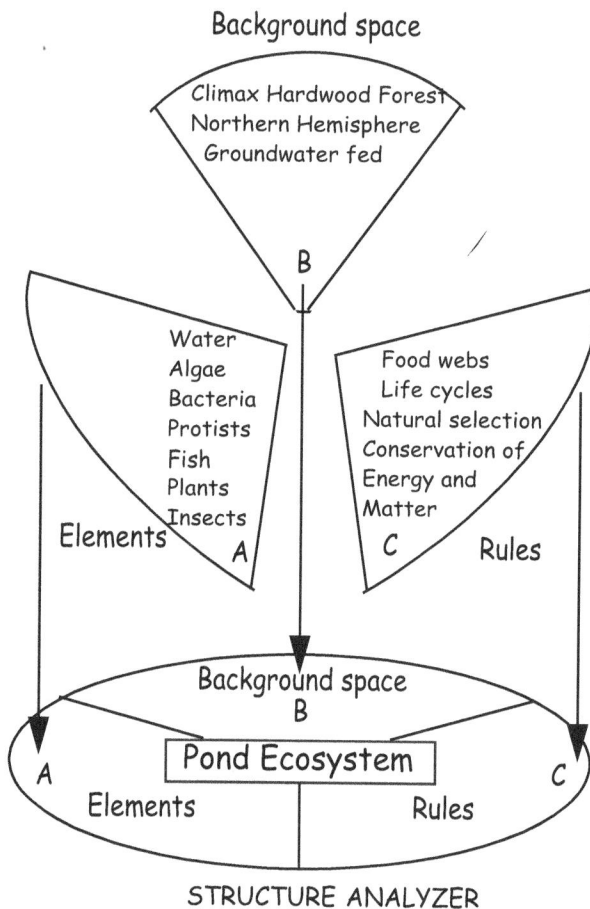

Background space

Climax Hardwood Forest
Northern Hemisphere
Groundwater fed

B

Water
Algae
Bacteria
Protists
Fish
Plants
Insects

Elements A

Food webs
Life cycles
Natural selection
Conservation of
Energy and
Matter

C Rules

Background space
B

Pond Ecosystem

A
Elements

C
Rules

STRUCTURE ANALYZER

Therefore, the force is directed at the element of water, which is reduced in the system.

The Change Analyzer will change over time, determined by the relationship of water to the arrangements between the living and non-living elements. If we suppose that decreased water will reduce the survivability of the algae and bacteria in the pond, then we might expect the food chains to fail to support the protists, fish, and insects that eat the plants. The plants might overgrow the pond (they will probably have enough water) and lead it toward a marsh, which then might become filled in first with wetland ferns and rushes, then locusts, beeches, birches, and finally, over time, maples and oaks.

The Change Analyzer might show these multiple stages as the pond evolves out of existence and into a portion of wetland climax forest, all due to the drought of 1999.

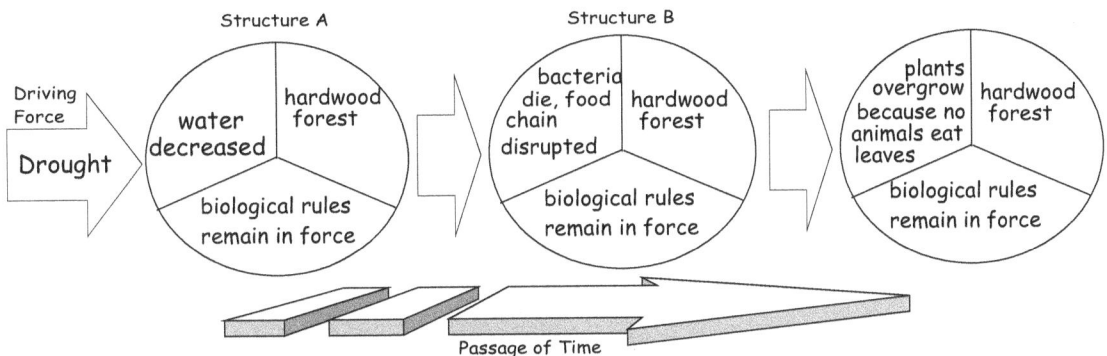

EVOLUTION AND CHANGE ANALYZERS

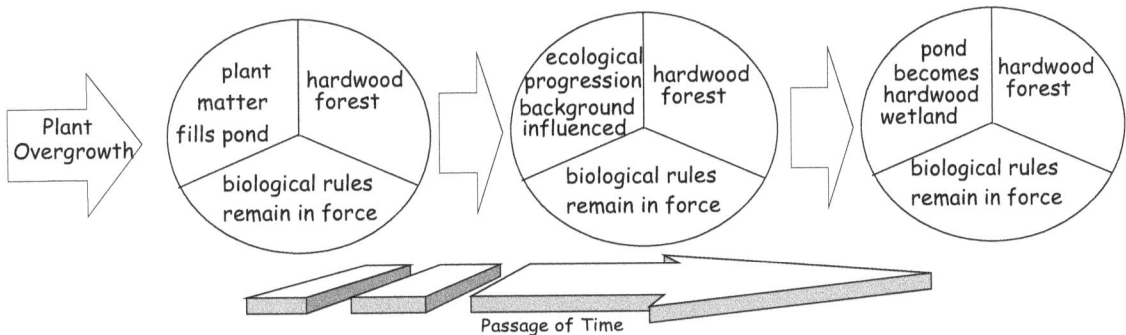

Cross-
Curricular
Example:
A Story Map

In this example, the Change and Evolution Analyzer acts as a standard Story Map, using a traditional reading/writing school exercise. The process not only extracts the story but allows a complete critical analysis to be immediately composed.

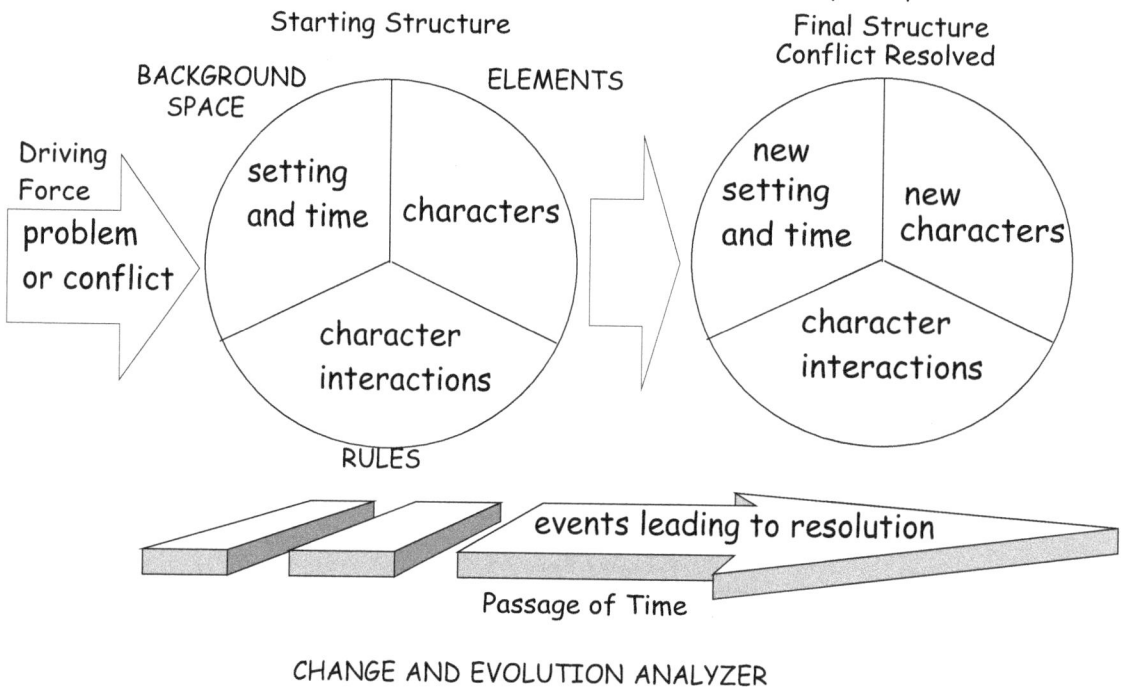

CHANGE AND EVOLUTION ANALYZER

CHAPTER 13

Analysis of More Complex Systems

Analyzers	Application
Compare and Contrast Analyzer	· Structures analysis of comparison between systems
Conflict Analyzer	· Defines conflicts between two observers of the same system

Compare and Contrast Analyzer

An attribute Compare and Contrast Analysis is an observable/feature by observable/feature sorting of two systems into a Sorting Analyzer. There are two (or more) independent systems being sorted by **the same observable** in a Compare and Contrast Analyzer. The operation of comparing and contrasting attributes is best viewed as a pair of side-by-side sorting keys with a shared bin into which are placed the common observable properties. The process is a sequential one in which an element from the sorting bin on the left is compared sequentially to each element from the sorting bin on the right.

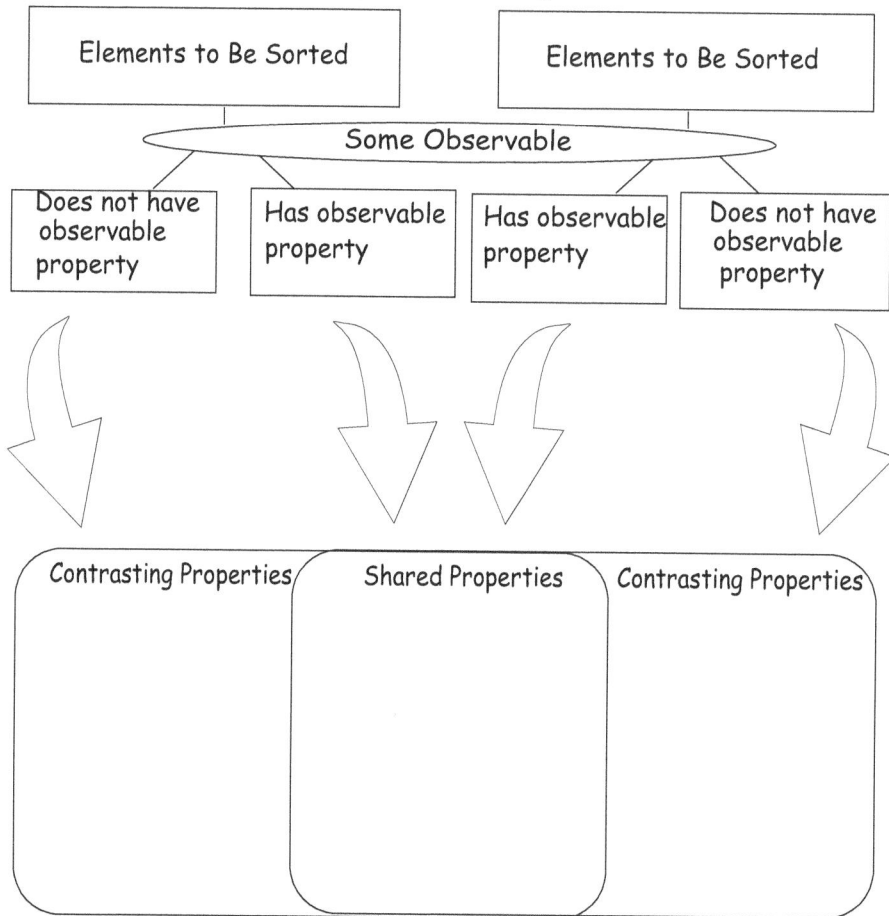

COMPARE AND CONTRACT ANALYZER

The process of comparing features in this way allows comparison, categorization, and identification of the differences and similarities between systems. This is an important step to finding connections and relationships that can be modeled and tested.

Other Compare and Contrast Analyzers are used in the Graphical Analyzer System. Often an important test is to compare the separate properties of two systems and observe if they are the "same". If all of the properties are the same, then the two systems are judged to be equivalent.

Practice Exercise

Compare and contrast the elements in these systems of geometric figures (answer next page).

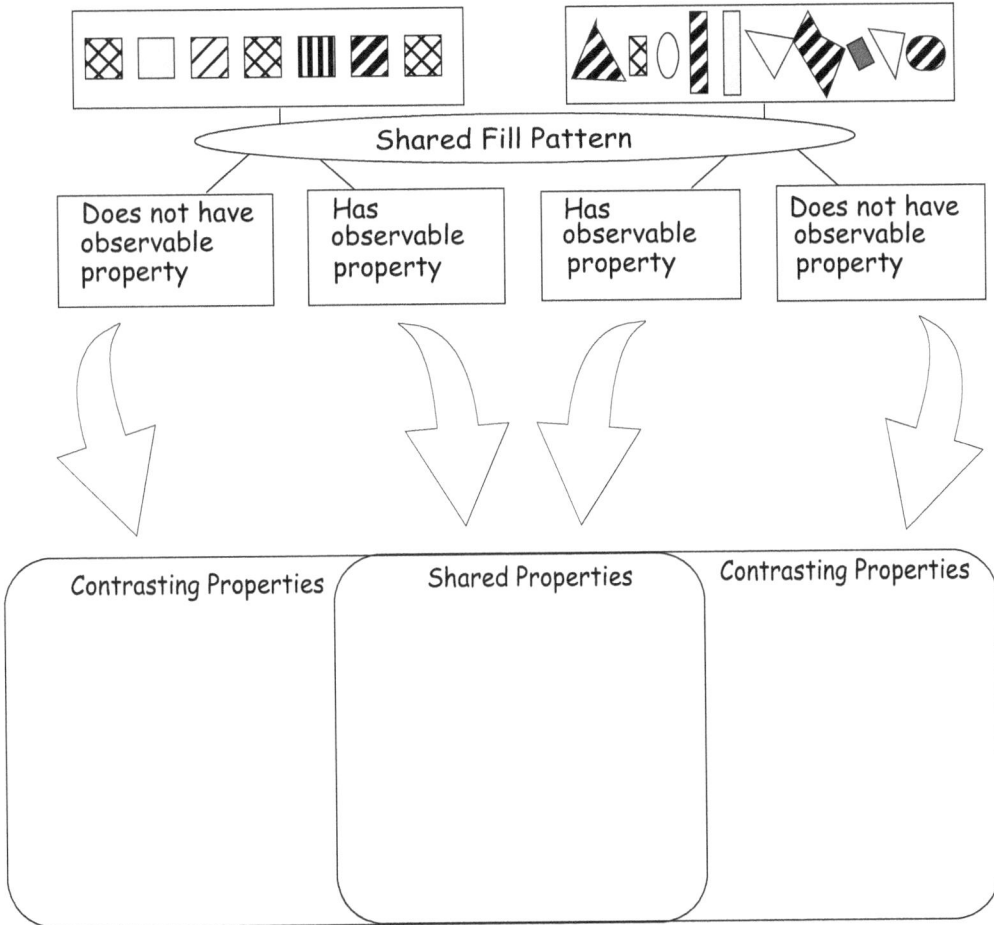

Shared Fill Pattern

| Does not have observable property | Has observable property | Has observable property | Does not have observable property |

Contrasting Properties | Shared Properties | Contrasting Properties

COMPARE AND CONTRAST ANALYZER

Answer

Answer to previous page.

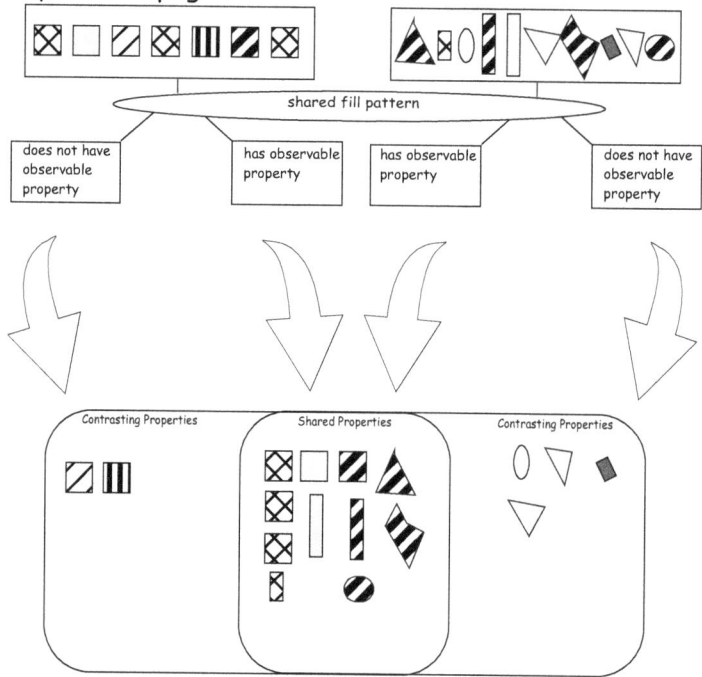

COMPARE AND CONTRAST ANALYZER

Practice
Exercise

Compare and contrast the properties associated with electrical and magnetic forces. The mental process tests each property on the left with each on the right in sequence.

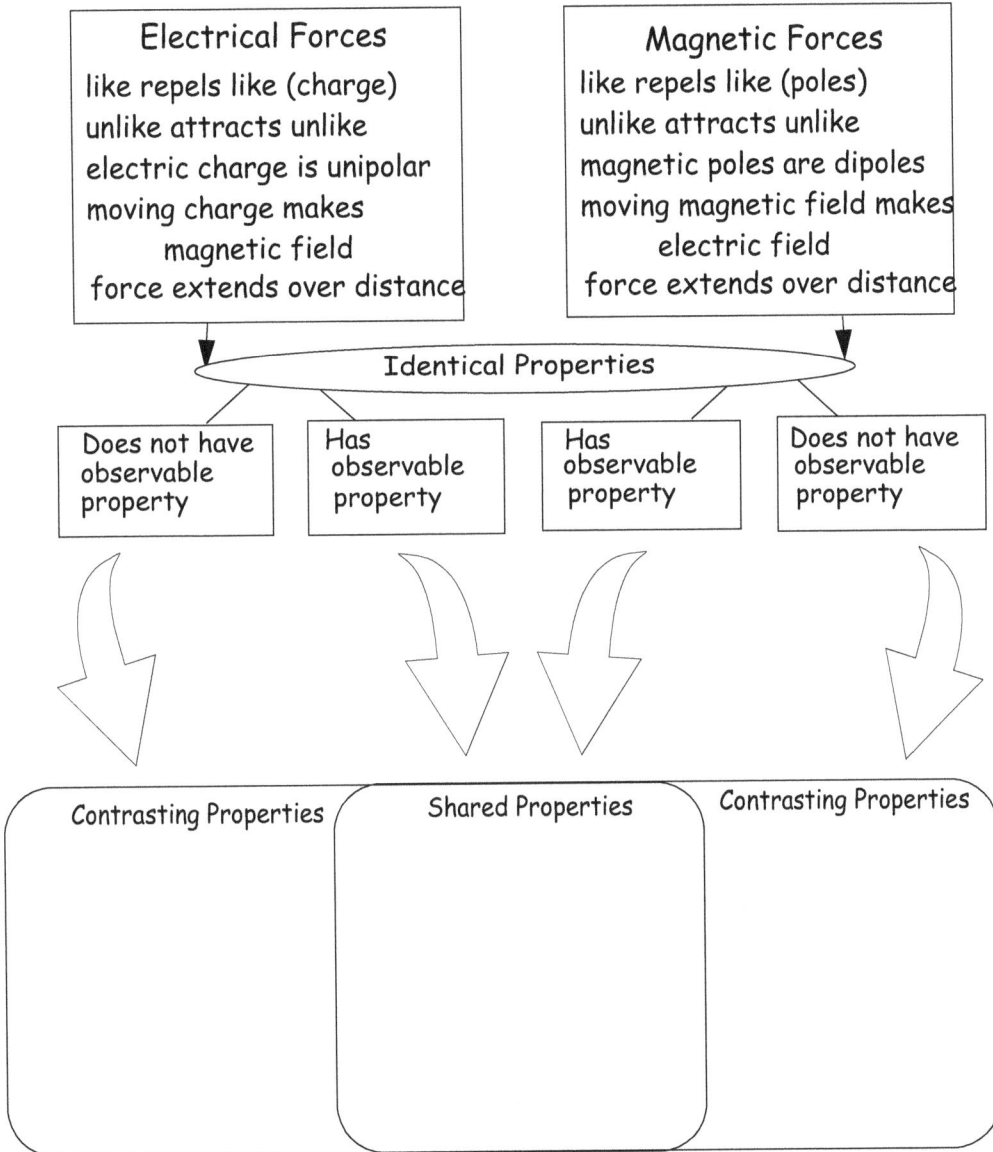

Electrical Forces	Magnetic Forces
like repels like (charge)	like repels like (poles)
unlike attracts unlike	unlike attracts unlike
electric charge is unipolar	magnetic poles are dipoles
moving charge makes magnetic field	moving magnetic field makes electric field
force extends over distance	force extends over distance

Identical Properties

Does not have observable property	Has observable property	Has observable property	Does not have observable property

Contrasting Properties	Shared Properties	Contrasting Properties

COMPARE AND CONTRAST ANALYZER

Answer   Answer to magnetic and electric force comparison.

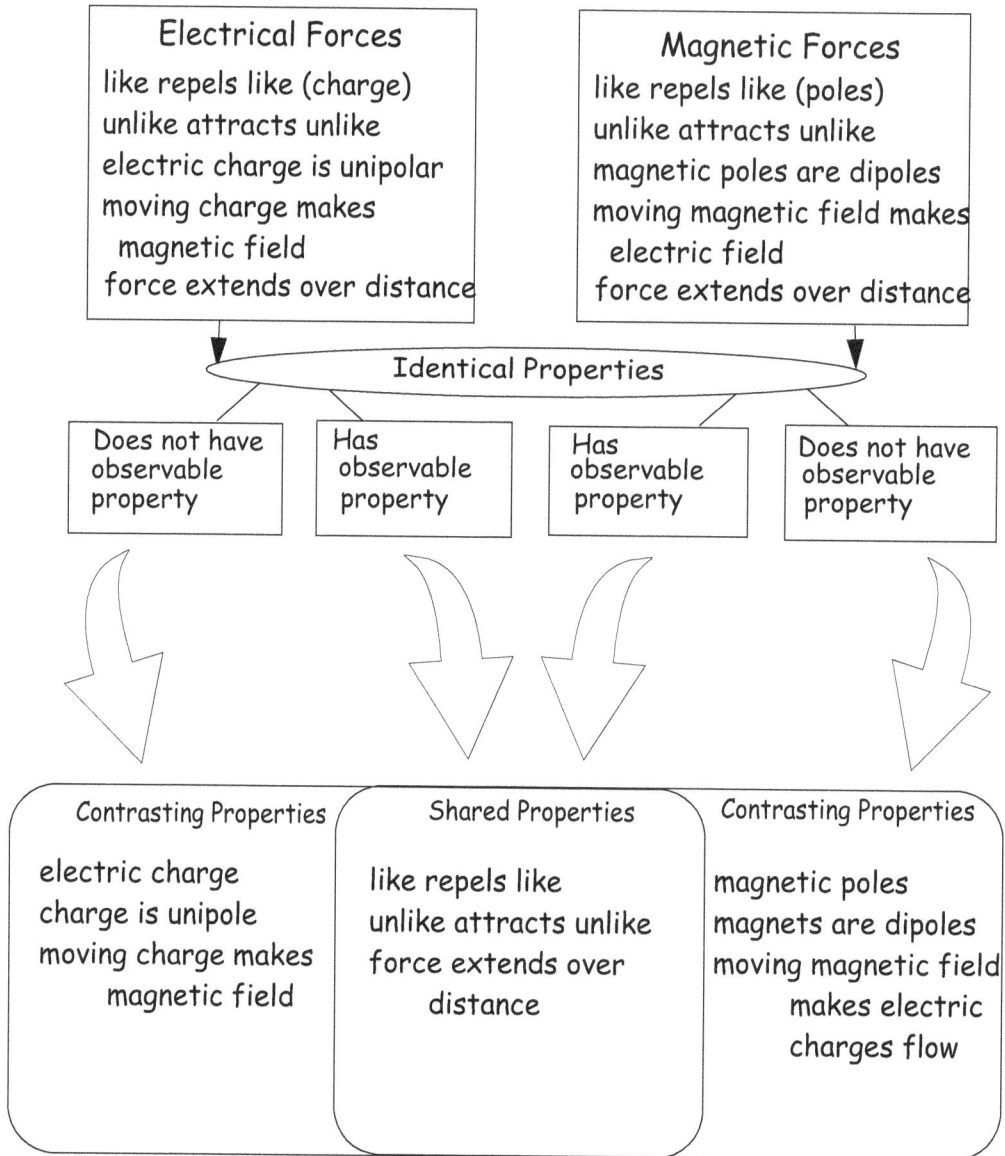

Electrical Forces	Magnetic Forces
like repels like (charge)	like repels like (poles)
unlike attracts unlike	unlike attracts unlike
electric charge is unipolar	magnetic poles are dipoles
moving charge makes magnetic field	moving magnetic field makes electric field
force extends over distance	force extends over distance

Identical Properties

Does not have observable property	Has observable property	Has observable property	Does not have observable property

Contrasting Properties	Shared Properties	Contrasting Properties
electric charge	like repels like	magnetic poles
charge is unipole	unlike attracts unlike	magnets are dipoles
moving charge makes magnetic field	force extends over distance	moving magnetic field makes electric charges flow

COMPARE AND CONTRAST ANALYZER

Conflict Analyzer

The Compare and Contrast Analyzer can be modified to a Conflict Analyzer to help discover the causes of disputes and conflicts. This is an important troubleshooting technique in everything from computer programming and car troubles to medical diagnosis and marriage counseling! In this Conflict Analyzer, a Venn-type diagram is combined with the Sorting Analyzer. The areas of overlap are areas of conflict. It is the observable that causes a conflicting view of the same system by two different observers.

Cross-Curricular Example

Here is an example that relates to both current events and to disputes in the classroom. Both John and Jim perceive that a particular space is theirs. The Conflict Analyzer will show the conflict to be in John's and Jim's choice of observable.

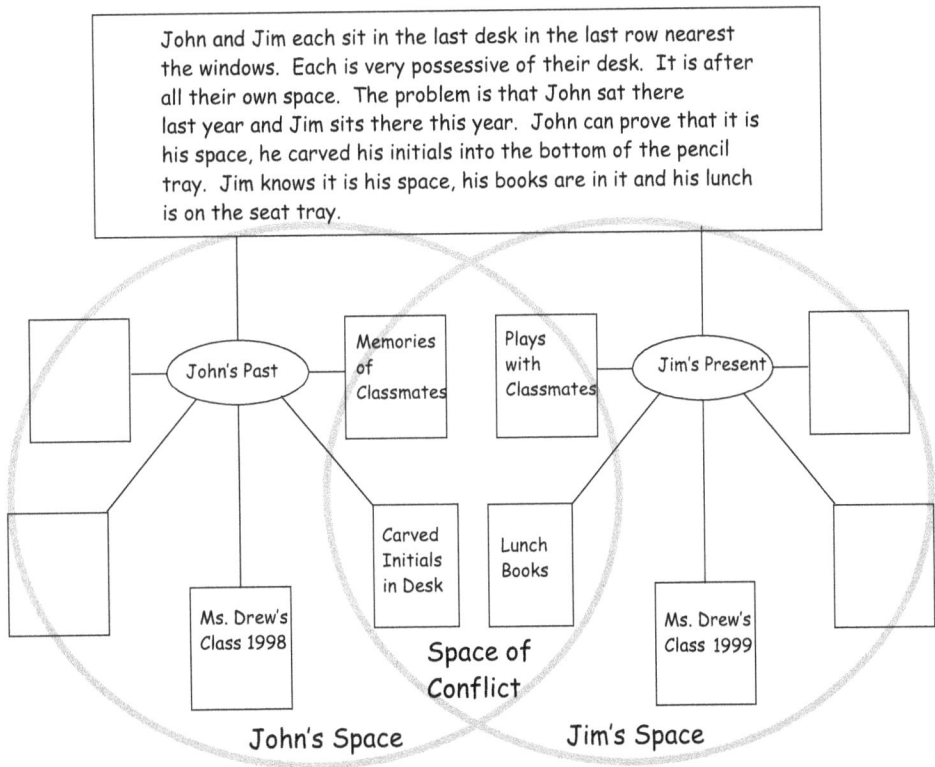

John and Jim each sit in the last desk in the last row nearest the windows. Each is very possessive of their desk. It is after all their own space. The problem is that John sat there last year and Jim sits there this year. John can prove that it is his space, he carved his initials into the bottom of the pencil tray. Jim knows it is his space, his books are in it and his lunch is on the seat tray.

John's Past — Memories of Classmates — Plays with Classmates — Jim's Present

Ms. Drew's Class 1998

Carved Initials in Desk — Lunch Books

Ms. Drew's Class 1999

Space of Conflict

John's Space   Jim's Space

Notes

CHAPTER 14

The Road to Understanding

This exploration of the Progression of Inquiry and Systems Analysis using the Language of Patterns and the Graphical Analyzer System is now complete. Its purpose has been to provide the tools needed to move you onto the road to the Age of Knowledge.

There are many more applications and a whole world that can be explored, described, measured, and known through the Language of Patterns and the Graphical Analyzer System. The tour is ended. Your journey begins.

For more information about
SymmetryScience materials

Symmetry Learning Systems

www.symmetrylearning.com

SYMMETRY LEARNING
PRESS
A subsidiary of
SYMMETRY RESEARCH, INC